Lecture Notes in Mathematics

A collection of informal reports and seminars

Edited by A. Dold, Heidelberg and B. Eckmann, Zürich

199

Charles J. Mozzochi

Yale University, New Haven, CT/USA

On the Pointwise Convergence of Fourier Series

Springer-Verlag

Berlin · Heidelberg · New York 1971

AMS Subject Classifications (1970): 43A50

ISBN 3-540-05475-8 Springer-Verlag Berlin · Heidelberg · New York
ISBN 0-387-05475-8 Springer-Verlag New York · Heidelberg · Berlin

Offsetdruck: Julius Beltz, Hemsbach

Dedicated to the memory of my father and mother

Foreword

This monograph is a detailed (essentially) self-contained treatment of the work of Carleson and Hunt and others needed to establish the Main Theorem: If $f \in L^p(-\pi,\pi)$ $1<p\leq\infty$, then $S_n(x;f)$ converges to $f(x)$ almost everywhere where $S_n(x;f)$ is the nth partial sum of the Fourier series for f.

The purpose of the first five chapters of this monograph is to develop the machinery necessary to reduce the proof of the Main Theorem to the proof (given in Chapter 6) of a theorem that involves only characteristic functions in $L^\infty(-4\pi,4\pi)$.

In Chapter 6 a large number of statements are presented without proof. Every proof that is omitted in Chapter 6 can be found in Appendix B.

The reader who is not familiar with the concept of Cauchy principal value (denoted: P.V.) or the concept of the Hilbert transform is referred to Appendix A for a summary of the results needed in the sequel.

Appendix C contains the recent results of Kahane and Katznelson on divergence sets which in a sense are opposite to those of Carleson and Hunt.

I would like to thank Professor S.A. Gaal, Professor R.E. Edwards, and Professor S. Kakutani for their encouragement during the preliminary stages of the writing of this monograph, and Professor R.A. Hunt for his very generous help during the spring and summer of 1969 at which time he explained in detail a portion of his original papers to me and made a number of suggestions for improving a few of the proofs contained in them.

Without his offer of assistance in the spring of 1969 I would not have seriously considered writing this monograph. Also, I would like to thank Professor Y. Katznelson for his permission to reproduce in Appendix C a portion of his text: An Introduction to Harmonic Analysis.

An outline of the proof of Theorem (2.2) based on Theorem (1.18) and Theorem (3.6) was communicated to me by Professor E.M. Stein and my appreciation for his doing so is to be noted here.

This paper was completed while the author was conducting a seminar on the contents of a preliminary draft at Yale University during the fall of 1969. I would like to further acknowledge my indebtedness to the men who participated in this seminar: James Arthur, Eugene J. Boyer, G.I. Gaudry, Michael Keane, S.J. Sidney, Joseph E. Sommese, Charles Stanton, A. Figa-Talamanca and Professors S. Kakutani and C.E. Rickart.

September, 1970 C. J. Mozzochi

TABLE OF CONTENTS

I. A THEOREM OF STEIN AND WEISS

Throughout this monograph we assume each function f is real-valued and in $L^1(-\pi,\pi)$ and hence finite almost everywhere in $(-\pi,\pi)$. We say f and g are equivalent iff $f(x) = g(x)$ for almost every x in $(-\pi,\pi)$. Hence we may assume, when necessary, that f is finite everywhere in $(-\pi,\pi)$.

The proofs in this chapter have been taken directly with only slight modification from [14].

m or μ denotes the Lebesgue measure on $(-\pi,\pi)$.

(1.1) Definition. For each $y > 0$ the function

$$\lambda_f(y) = m\{x \in (-\pi,\pi) \mid |f(x)| > y\}$$

is called the distribution function of f.

(1.2) Remark. Since $\lambda_f(y) < \infty$ for each $y > 0$ and f is finite almost everywhere $\lim_{y\to\infty} \lambda_f(y) = 0$. Clearly, λ_f is non-negative and non-increasing.

Using the fact

$$\bigcup_{n=1}^{\infty} \{x \in (-\pi,\pi) \mid |f(x)| > y_0 + \frac{1}{n}\} = \{x \in (-\pi,\pi) \mid |f(x)| > y_0\}$$

we have that λ_f is continuous from the right. Since λ_f is monotonic, it has a countable number of discontinuities; so that it is measurable.

Let T be a mapping from a subset of the integrable real-valued functions defined on $(-\pi,\pi)$ that contains the simple functions into the set of measurable real-valued functions defined on $(-\pi,\pi)$. In this chapter we assume $1 < p < \infty$, $1 < q < \infty$.

(1.3) Definition. T is of type (p,q) iff there exists $A > 0$ such that $\|Tf\|_q \le A\|f\|_p$ for every simple function f.

(1.4) Definition. T is of weak type (p,q) iff there exists $A > 0$ such that for each simple function f and $y > 0$

$$\lambda_{Tf}(y) \leq \left[\frac{A}{y} \|f\|_p\right]^q$$

(1.5) Definition. T is of restricted type (p,q) iff there exists $A > 0$ such that for each measurable set $E \subset (-\pi,\pi)$

$\|T\chi_E\|_q \leq A \; \|\chi_E\|_p$, where χ_E is the characteristic function of E.

(1.6) Definition. T is of restricted weak type (p,q) iff there exists $A > 0$ such that for each measurable set $E \subset (-\pi,\pi)$

$$\lambda_{T\chi_E}(y) \leq \left[\frac{A}{y} \|\chi_E\|_p\right]^q$$

(1.7) Lemma. If T is of restricted type (p,q), then T is of restricted weak type (p,q).

Proof. Let $E_y = \{x \in (-\pi,\pi) \mid |T\chi_E(x)| > y\}$. Let E be any measurable set contained in $(-\pi,\pi)$.

$$(A\|\chi_E\|_p)^q \geq \|T\chi_E\|_q^q = \int_{-\pi}^{\pi} |T\chi_E|^q \, d\mu \geq$$

$$\int_{E_y} |T\chi_E|^q d\mu \geq y^q \int_{E_y} d\mu = y^q \mu(Ey) = y^q \lambda_{T\chi_E}(y)$$

(1.8) Lemma. For $1 \leq p < \infty$ and $f \in L^1(-\pi,\pi)$ we have

$$\int_{-\pi}^{\pi} |f|^p d\mu = \int_0^\infty p y^{p-1} \lambda_f(y) \; dy$$

Proof.

$$\int_{-\pi}^{\pi} |f|^p d\mu = \int_{-\pi}^{\pi} \left(\int_0^{|f(x)|} py^{p-1} dy \right) d\mu = \int_{-\pi}^{\pi} \left(\int_0^{\infty} \chi_{[0,|f(x)|)}(y)\; py^{p-1} dy \right) d\mu ;$$

so that by Fubini's theorem since the set $\{(x,y) \mid x \in (-\pi,\pi)$ and $|f(x)| > y\}$ is product measurable

$$\int_{-\pi}^{\pi} |f|^p d\mu = \int_0^{\infty} \left(py^{p-1} \int_{-\pi}^{\pi} \chi_{[0,|f(x)|)}(y)\, d\mu \right) dy. \quad \text{But}$$

$$\chi_{[0,|f(x)|)}(y) = \chi_{E_y}(x) \quad \text{where } E_y = \{x \in X \mid |f(x)| > y\}.$$

This completes the proof of (1.8).

In the rest of this chapter we assume:

$1 < p_k \leq q_k < \infty$ $(k = 0,1)$, $p_0 \neq p_1$, $q_0 \neq q_1$ and for $0 < t < 1$

$$\frac{1}{p_t} = \frac{(1-t)}{p_0} + \frac{t}{p_1} \; ; \quad \frac{1}{q_t} = \frac{(1-t)}{q_0} + \frac{t}{q_1}$$

If $s \geq 1$, then s' is that number (including ∞) satisfying $\left(\frac{1}{s} + \frac{1}{s'} \right) = 1$

(1.9) <u>Lemma.</u> Let T be of restricted weak type (p_0, q_0) (p_1,q_1), then it is of restricted type (p_t,q_t) for $0 < t < 1$.

<u>Proof.</u> Let $p = p_t$, $q = q_t$ for a fixed t between 0 and 1. Suppose $E \subset (-\pi,\pi)$ is a measurable set, χ_E its characteristic function, $h = T\chi_E$ and $\lambda(y)$ the distribution function of h. We can assume, without loss of generality, that $q_0 < q_1$. Then, using the restricted weak type relations (with constants A_0 and A_1),

we obtain by (1.8) for any $C > 0$

$$\int_{-\pi}^{\pi} |h|^q d\mu = q\int_0^{\infty} y^{q-1}\lambda(y)\,dy = q\int_0^{C} y^{q-1}\lambda(y)\,dy + q\int_C^{\infty} y^{q-1}\lambda(y)\,dy$$

$$\leq q\int_0^{C} y^{q-1}\left(\frac{A_o}{y}[\mu(E)]^{1/p_o}\right)^{q_o} dy + q\int_C^{\infty} y^{q-1}\left(\frac{A_1}{y}[u(E)]^{1/p_1}\right)^{q_1} dy$$

$$= \left[q\,\frac{A_o^{q_o}}{(q-q_o)}\right][\mu(E)]^{q_o/p_o} C^{q-q_o} + \left[q\,\frac{A_1^{q_1}}{(q_1-q)}\right][\mu(E)]^{q_1/p_1} C^{q-q_1}$$

Letting $C = [\mu(E)]^s$, where

$$s = \frac{1}{(q-q_o)}\left(\frac{q}{p} - \frac{q_o}{p_o}\right) = \frac{1}{(q-q_1)}\left(\frac{q}{p} - \frac{q_1}{p_1}\right)$$

we have

$$[\mu(E)]^{q_o/p_o} C^{(q-q_o)} = [\mu(E)]^{q/p} = [\mu(E)]^{q_1/p_1} C^{q-q_1}$$

Thus we have shown that

$$\int_{-\pi}^{\pi} |h|^q d\mu \leq A^q [\mu(E)]^{q/p},$$

where

$$A = \left(\frac{q\,A_o^{q_o}}{q-q_o} + \frac{q\,A_1^{q_1}}{q_1-q}\right)^{1/q}$$

This completes the proof of (1.9)

Suppose T is linear and of restricted type (p,q) and let $f \in L^{q'}(-\pi,\pi)$. Let γ be the set-function, defined on the measurable subsets $E \subset (-\pi,\pi)$, such that

$$(1.10)\qquad \gamma(E) = \int_{-\pi}^{\pi} (T\chi_E) f\,d\mu$$

Since T is of restricted type (p,q), γ is countably additive and absolutely continuous with respect to μ . Thus by the Radon-Nikodym theorem, there exists a unique (almost everywhere) function h on $(-\pi,\pi)$, such that

$$\gamma(E) = \int_E h\, d\mu \tag{1.11}$$

for each measurable set $E \subset (-\pi,\pi)$. Define the operator T^*, acting on $L^{q'}(-\pi,\pi)$, by letting $T^*f = h$. T^* is clearly linear; also, it behaves, at least formally, like the adjoint operator to T. That is, if s is a real-valued simple function defined on $(-\pi,\pi)$ and f is in $L^{q'}(-\pi,\pi)$, we obtain from (1.10) and (1.11) and the linearity of T and T^*

$$\int_{-\pi}^{\pi} (Ts)\, f\, d\mu = \int_{-\pi}^{\pi} s(T^*f)d\mu. \tag{1.12}$$

In general, however, it is not true that T^*f is in $L^{p'}(-\pi,\pi)$ for all f in $L^{q'}(-\pi,\pi)$.

(1.13) Lemma. Suppose T is linear and of restricted type (p,q), where $1 < p < \infty$ and $1 \leq q < \infty$, then T^* is of weak type (q',p').

Proof. Let f be in $L^{q'}(-\pi,\pi)$ and $h = T^*f$. If $E_y = \{x \in (-\pi,\pi) \mid |h(x)| > y\}$ and $\lambda(y) = \mu(E_y)$ is the distribution function of h, we must show the existence of a positive number B, independent of f in $L^{q'}(-\pi,\pi)$, satisfying for each $y > 0$

$$\lambda(y) \leq \left[\frac{B}{y} \|f\|_{q'}\right]^{p'} \tag{1.14}$$

Actually, according to our definitions we need only consider T^* restricted to the class of real-valued simple functions on $(-\pi,\pi)$, but the proof of (1.14) will yield the result for all functions in $L^{q'}(-\pi,\pi)$. We put $E_y = E_y^+ \cup E_y^-$, where $E_y^+ = \{x \in (-\pi,\pi) \mid h(x) > y\}$

$E^-_y = \{x \in (-\pi,\pi) \mid h(x) < -y\}$. Let $\lambda^+(y) = \mu(E^+_y)$, $\lambda^-(y) = \mu(E^-_y)$. Then, for $y > 0$, $E^+_y \cap E^-_y = 0$ and $\lambda(y) = \mu(Ey) = \mu(E^+_y) + \mu(E^-_y) = \lambda^+(y) + \lambda^-(y)$. Thus by the definition of $\lambda^+(y)$, (1.12) with $s = \chi_{E_y^+}$, Holder's inequality and the assumption of restricted type (p,q) we have

$$y\ \lambda^+(y) = y \int_{E_y^+} d\mu \leq \int_{E_y^+} h\, d\mu = \int_{-\pi}^{\pi} \chi_{E_y^+} (T^*f) d\mu = \int_{-\pi}^{\pi} (T\chi_{E_y^+}) f\, d\mu$$

$$\leq \|T\chi_{E_y^+}\|_q \cdot \|f\|_{q'} \leq A \|\chi_{E_y^+}\|_p \cdot \|f\|_{q'} .$$

Since $\|\chi_{E_y^+}\|_p = [\lambda^+(y)]^{1/p}$, we have shown, for $y > 0$,

$$y\ \lambda^+(y) \leq A[\lambda^+(y)]^{1/p} \|f\|_{q'} .$$

But $1 - (1/p) = (1/p')$; so that this last inequality can be transformed into

$$(1.15) \qquad \lambda^+(y) \leq \left[\frac{A}{y} \|f\|_{q'}\right]^{p'} \text{ for } y > 0.$$

By a completely analogous argument we obtain, for $y > 0$,

$$(1.16) \qquad \lambda^-(y) \leq \left[\frac{A}{y} \|f\|_{q'}\right]^{p'} .$$

Since $\lambda(y) = \lambda^+(y) + \lambda^-(y)$, (1.15) and (1.16) yield (1.14) with $B = 2^{1/p'} A$.

(1.17) Theorem. (Marcinkiewicz). If T is linear and of weak types (p_o, q_o) and (p_1, q_1), then it is of type (p_t, q_t).

Proof. Cf. [16] Vol II chapter XII.

(1.18) Theorem. (Stein-Weiss). If T is linear and of restricted weak type (p_o,q_o) and (p_1,q_1), then it is of type (p_t,q_t).

Proof. Since T is of restricted weak types (p_o,q_o) and (p_1,q_1), by (1.9) it is of restricted type (p_t,q_t) for $0 < t < 1$.
Hence by (1.13) T* is of weak type (q'_t, p'_t). Suppose $0 < t_o < s < t_1 < 1$. The conditions $p_o \neq p_1$, $1 < p_k \le q_k$ $(k = 0,1)$ imply $p'_o \neq p'_1$, $1 \le q'_k \le p'_k$ $(k = 0,1)$. Thus we have $p'_{t_o} \neq p'_{t_1}$, $1 < q'_{t_k} \le p'_{t_k}$ $(k = 0,1)$. Hence we can apply (1.17) to T* with (p_k,q_k) replaced by (q'_{t_k}, p'_{t_k}) $(k = 0.1)$, obtaining the result that T* is of type (q'_s, p'_s). It follows that T* is defined on all of $L^{q'_s}(-\pi,\pi)$ and is a bounded operator into $L^{p'_s}(-\pi,\pi)$. Thus the adjoint of T* is also bounded. But it follows from (1.12) and the representation theorem of F. Riesz that this adjoint must agree with T on the class of all real-valued simple functions defined on $(-\pi,\pi)$. Thus T is of type (p_s,q_s). By making t_o tend to 0 and t_1 to 1, if necessary, we see that s can be any number satisfying $0 < s < 1$.

(1.19) Remark. It is very important for the application of (1.18) in Chapter III that the reader convince himself that the bounds of the operator in (1.18) (as well as in (1.17)) as a transformation on $L^{p_t}(-\pi,\pi)$ into $L^{q_t}(-\pi,\pi)$ depend only on t, p_o, q_o, p_1, q_1 and the constants of the weak type (p_o,q_o) and (p_1,q_1) inequalities.

II. THE MAIN THEOREM

Let $f \in L^1(-\pi,\pi)$. Let $S_n(x;f)$ be the nth partial sum of the Fourier series for f. Let $1 < p < \infty$.

Let $M : L^p(-\pi,\pi) \rightarrow$ class of extended real valued functions on $(-\pi,\pi)$.

$$Mf : (-\pi,\pi) \rightarrow [o,\infty]$$

$$Mf(x) = \sup \{ |S_n(x;f)| \, | n \geq o \}$$

(2.1) <u>Remark.</u> It is easily shown that M is sublinear.

In chapter 3 we give a proof of the following

(2.2) <u>Theorem.</u> For every $f \in L^p(-\pi,\pi)$ $1 < p < \infty$

$$\|Mf\|_p \leq C_p \|f\|_p$$

where $C_p > 0$ is a constant dependent on p but independent of f.

(2.3) <u>Remark.</u> The reader should recall the following well known facts about L^p spaces: (A) If $1 \leq p \leq \infty$ and $\|f_n - f\|_p \rightarrow o$, then there exists a subsequence $\{n_k\}$ such that $f_{n_k}(x) \rightarrow f(x)$ almost everywhere. (B) For every $f \in L^p(-\pi,\pi)$ $1 \leq p < \infty$ and for every $\varepsilon > 0$ there exists a polynomial $P \in L^p(-\pi,\pi)$ such that $\|f-P\|_p < \varepsilon$. (C) If $1 \leq p \leq \infty$; $1 \leq q \leq \infty$, then $p > q$ implies $L^q(-\pi,\pi) \supseteq L^p(-\pi,\pi)$.

(2.4) <u>Main</u> <u>Theorem</u> (Carleson-Hunt) If $f \in L^p(-\pi,\pi)$ $1 < p \leq \infty$, then $S_n(x;f) \rightarrow f(x)$ for almost every x in $(-\pi,\pi)$.

(2.5) Lemma. If $f \in L^p(-\pi,\pi)$ $1 \leq p < \infty$, then for every $0 < \varepsilon < 1$ there exists a sequence $\{\varepsilon_k\}$ of positive real numbers and a sequence $\{P_k\}$ of polynomials in $L^p(-\pi,\pi)$ such that $0 < \varepsilon_k^p < \varepsilon^k$ for all k, $\varepsilon_k \to 0$, $P_k(x) \to f(x)$ for almost every x in $(-\pi,\pi)$, and $\| f-P_k \|_p < \varepsilon_k^2$ for all k.

Proof. This is immediate by (2.3) (A) and (2.3) (B).

(2.6) Lemma. For each k let $E_k = \{x \in (-\pi,\pi) \mid M(f-P_k)(x) > \varepsilon_k\}$ where ε_k and P_k are those constructed in (2.5). Then for $1 < p < \infty$ we have $m\, E_k \leq C_p^p \, \varepsilon_k^p$ where C_p is that of (2.2) and $f \in L^p(-\pi,\pi)$.

Proof. $\varepsilon_k^p \, m(E_k) = \int_{E_k} \varepsilon_k^p \, dx \leq \int_{E_k} \left[M(f-P_k)(x) \right]^p dx \leq$

$$\int_{-\pi}^{\pi} [M(f-P_k)(x)]^p \, dx = \| M(f-P_k) \|_p^p \leq C_p^p \, \| f-P_k \|_p^p \text{ by (2.2).}$$

But by (2.5) we have $\| f-P_k \|_p^p \leq \varepsilon_k^{2p}$.

Proof of Theorem (2.4)

Clearly, by (2.3) (C) it is sufficient to show that $S_n(x;f) \to f(x)$ for almost every x in $(-\pi,\pi)$ if $1 < p < \infty$. It is easy to show that for each k $S_n(x;f) = S_n(x; f-P_k) + S_n(x;P_k)$ for all n and for all $x \in (-\pi,\pi)$; so that for each k, for each n and for all $x \in (-\pi,\pi)$ we have $|S_n(x;f) - f(x)| < |S_n(x;f-P_k)| + |S_n(x;P_k) - f(x)|$. But since $P_k(x)$ is differentiable on $(-\pi,\pi)$ for each k we have for each k $\lim_{n\to\infty} S_n(x;P_k) = P_k(x)$ for each x in $(-\pi,\pi)$. Hence

for each k and for each x in $(-\pi,\pi)$

$$\overline{\lim_{n\to\infty}} |S_n(x;f) - f(x)| \le \overline{\lim_{n\to\infty}} |S_n(x;f-P_k)| + |f(x) - P_k(x)| .$$

But for each k and for each x in $(-\pi,\pi)$

$$\overline{\lim_{n\to\infty}} |S_n(x;f-P_k)| \le M(f-P_k)(x)$$

Hence for each k and for each x in $(-\pi,\pi)$

$$\overline{\lim_{n\to\infty}} |S_n(x;f) - f(x)| \le M(f-P_k)(x) + |f(x) - P_k(x)| .$$

By (2.6) if $x \notin E_k$, then $M(f-P_k)(x) \le \varepsilon_k$ where $mE_k \le C_p^p \varepsilon_k^p$.

Consequently, $M(f-P_k)(x) \to 0$ for all $x \notin \bigcup_{k=1}^{\infty} E_k = A$.

But

$$m(A) \le \sum_{k=1}^{\infty} m(E_k) \le C_p^p \sum_{k=1}^{\infty} \varepsilon_k^p \le C_p^p \sum_{k=1}^{\infty} \varepsilon^k = C_p^p\left(\frac{\varepsilon}{1-\varepsilon}\right) \quad \text{since } \varepsilon <$$

And

$$|f(x) - P_k(x)| \to 0 \quad \text{for } x \text{ in } (-\pi,\pi) \text{ and } x \notin B$$

where $m(B) = 0$. Consequently,

$$\lim_{n\to\infty} S_n(x;f) = f(x) \text{ for } x \text{ in } (-\pi,\pi) \text{ but } x \notin (A\cup B) \quad \text{where}$$

$$m(A\cup B) \le C_p^p\left(\frac{\varepsilon}{1-\varepsilon}\right).$$

III. A PROOF OF THEOREM (2.2)

For every $f \in L^1(-\pi,\pi)$ let f^o denote its 2π-periodic extension with domain $(-4\pi,4\pi)$.

For every $f \in L^1(-\pi,\pi)$ and $x \in (-\pi,\pi)$

$$\text{let } S_n^*(x;f) = \text{P.V.} \int_{|x-t|<\pi} \frac{\varepsilon^{-int} f^o(t)}{x-t}\, dt \; ; \; |n| \geq 0.$$

$$\text{let } S_n^*(x;f;\omega^*_{-1}) = \text{P.V.} \int_{-4\pi}^{4\pi} \frac{\varepsilon^{-int} f^o(t)}{x-t}\, dt \; ; \; |n| \geq 0.$$

For $1 < p < \infty$ let

$M^*: L^p(-\pi,\pi) \to$ class of extended real valued functions on $(-\pi,\pi)$

$M^*f : (-\pi,\pi) \to [0,\infty]$

$M^*f(x) = \sup \{ |S_n^*(x;f;\omega^*_{-1})| \mid |n| \geq 0\}$

(3.1) Remark. We introduce the integrals $S_n^*(x;f)$ and $S_n^*(x;f;\omega^*_{-1})$ so that in the sequel we can use the machinery associated with the Hilbert transform (cf. Appendix A). Note that $S_n^*(x;f)$ is the complex conjugate of $S_{-n}^*(x;f)$ for $|n| \geq 0$ and similarly for $S_n^*(x;f;\omega^*_{-1})$. Also, it is easily shown that M^* is sublinear.

Let

$$D_n(y) = \begin{cases} \dfrac{\sin(n+\frac{1}{2})y}{2\sin\frac{y}{2}} & \text{if } y \in [-\pi,\pi] - \{0\} \\ (n+\frac{1}{2}) & \text{if } y = 0 \end{cases}$$

$$F_n(y) = \begin{cases} \dfrac{\sin ny}{y} & \text{if } y \in [-\pi,\pi] - \{0\} \\ n & \text{if } y = 0 \end{cases}$$

$$G_n(y) = D_n(y) - F_n(y) \quad \text{for all } y \in [-\pi,\pi].$$

(3.2) Remark. It is clear that $D_n(y)$, $F_n(y)$ and $G_n(y)$ are continuous on $[-\pi,\pi]$.

(3.3) Lemma. $|G_n(y)| \leq C_1$ for all $n \geq 0$ and for all $y \in [-\pi,\pi]$ where $C_1 > 0$.

Proof. Let

$$g(y) = \begin{cases} \dfrac{1}{2\tan\frac{y}{2}} - \dfrac{1}{y} & \text{if } y \in [-\pi,\pi] - \{0\} \\ 0 & \text{if } y = 0 \end{cases}$$

By L'Hôpital's rule it is easily seen that $g(y)$ is continuous and bounded on $[-\pi,\pi]$.

But if $y \neq 0$ we have by direct calculation that

$$D_n(y) = \frac{\sin(n+\frac{1}{2})y}{2\sin\frac{y}{2}} = \frac{\sin ny \cos\frac{y}{2} + \cos ny \sin\frac{y}{2}}{2\sin\frac{y}{2}}$$

$$= \frac{\sin ny}{2\tan\frac{y}{2}} + \frac{1}{2}\cos ny$$

$$= \frac{\sin ny}{y} + g(y)\sin ny + \frac{1}{2}\cos ny = F_n(y) + g(y)\sin ny + \frac{1}{2}\cos ny,$$

And if $y = 0$, then by direct substitution we have

$D_n(y) = (n + \frac{1}{2}) = F_n(y) + \frac{1}{2}$

Consequently, for all $y \in [-\pi,\pi]$ we have that

$D_n(y) = F_n(y) + g(y)\sin ny + \frac{1}{2}\cos ny$; so that

$|D_n(y) - F_n(y)| \leq (|g(y)| + \frac{1}{2}) \leq C_1$ for all $n \geq 0$ and for all $y \in [-\pi,\pi]$.

(3.4) <u>Lemma</u> $Mf(x) \leq E_p(\|f\|_p + M^*f(x))$ for $1 < p < \infty$, for almost every x in $(-\pi,\pi)$ where $E_p > 0$ is a constant independent of f but dependent on p where $f \in L^p(-\pi,\pi)$.

<u>Proof.</u> For each $n \geq 0$ and $x \in (-\pi,\pi)$

$$S_n(x;f) = \frac{1}{\pi}\int_{-\pi}^{\pi} f^o(t)\, D_n(x-t)dt = \frac{1}{\pi}\int_{|x-t|<\pi} f^o(t)\, D_n(x-t)dt =$$

$$\frac{1}{\pi}\int_{|x-t|<\pi} f^o(t)\, G_n(x-t)dt + \frac{1}{\pi}\int_{|x-t|<\pi} f^o(t) F_n(x-t)dt$$

But for every $x \in (-\pi,\pi)$ we have

$$\frac{1}{\pi} \int_{|x-t|<\pi} f^{o}(t)F_n(x-t)dt = \frac{1}{\pi} \text{ P.V.} \int_{|x-t|<\pi} \frac{e^{in(x-t)} - e^{-in(x-t)}}{2i\ (x-t)} f^{o}(t)dt$$

But for almost every $x \in (-\pi,\pi)$ we have

$$\frac{1}{\pi} \text{ P.V.} \int_{|x-t|<\pi} \frac{e^{in(x-t)} - e^{-in(x-t)}}{2i\ (x-t)} f^{o}(t)dt = \frac{e^{inx}}{2\pi i} \text{ P.V.} \int_{|x-t|<\pi} \frac{e^{-int} f^{o}(t)}{x-t}dt -$$

$$\frac{e^{-inx}}{2\pi i} \text{ P.V.} \int_{|x-t|<\pi} \frac{e^{-i(-n)t} f^{o}(t)}{x-t} dt$$

$= \frac{e^{inx}}{2\pi i} S^*_n(x;f) - \frac{e^{-inx}}{2\pi i} S^*_{-n}(x;f)$; so that for almost every $x \in (-\pi,\pi)$

and for every $n \geq 0$ we have

$$|S_n(x;f)| \leq \frac{1}{\pi} \int_{|x-t|<\pi} |f^{o}(t)| \ |G_n(x-t)|\ dt + \frac{1}{2\pi}|S^*_n(x;f)| + \frac{1}{2\pi}|S^*_{-n}(x;f)|$$

Consequently, by (3.3) for $n \geq 0$ and for almost every $x \in (-\pi,\pi)$ we have $|S_n(x;f)| \leq \frac{C_1}{\pi}\|f\|_1 + \frac{1}{2\pi}(|S^*_n(x;f)| + |S^*_{-n}(x;f)|)$ and

since $|S^*_n(x;f)| = |S^*_{-n}(x;f)|$ we have for $n \geq 0$ and for almost every $x \in (-\pi,\pi)$ that

$|S_n(x;f)| \leq C_2 \|f\|_1 + C_2 |S^*_n(x;f)|$ where $C_2 > (C_1+1) \frac{1}{\pi} > 0$.

But for $n \geq 0$ and $x \in (-\pi,\pi)$ we have

$$S^*_n(x;f) = S^*_n(x;f;\omega^*_{-1}) - \int_{\substack{|x-t|\geq\pi \\ |t|<4\pi}} \frac{e^{-int} f^{o}(t)dt}{x-t}; \text{ so that}$$

$$|S^*_n(x;f)| \leq |S^*_n(x;f;\omega^*_{-1})| + \frac{4}{\pi}\|f\|_1$$

Consequently, for $n \geq 0$ and for almost every $x \in (-\pi,\pi)$

$$|S_n(x,f)| \leq C_2(1+\frac{4}{\pi}) \|f\|_1 + C_2 \,|S^*_n(x;f;\omega^*_{-1})|$$

But by Holder's inequality it is easily shown that

$$\|f\|_1 \leq (2\pi)^{(1-\frac{1}{p})}\|f\|_p \qquad 1 < p < \infty$$

Consequently, for $n \geq 0$ and for almost every $x \in (-\pi,\pi)$

$$|S_n(x;f)| \leq C_2 \,(1+\frac{4}{\pi})\,(2\pi)^{(1-\frac{1}{p})}\|f\|_p + C_2\,|S^*_n(x;f;\omega^*_{-1})|;\; n \geq 0$$

Let $E_p > \left(C_2(1+\frac{4}{\pi})\,(2\pi)^{(1-\frac{1}{p})} + C_2\right) > 0.$

Then for every $n \geq 0$ and for almost every $x \in (-\pi,\pi)$ we have

$$|S_n(x;f)| \leq E_p \,(\|f\|_p + |S^*_n(x;f;\omega^*_{-1})|)$$

But since $|S^*_n(x;f;\omega^*_{-1})| = |S^*_{-n}(x;f;\omega^*_{-1})|$ for every $n \geq 0$,

we have for almost every $x \in (-\pi,\pi)$

$$\sup_{n\geq o} |S_n(x;f)| \leq E_p \,(\|f\|_p + \sup_{|n|\geq o} |S^*_n(x;f;\omega^*_{-1})|)$$

so that for almost every $x \in (-\pi,\pi)$ we have

$$Mf(x) \leq E_p \,(\|f\|_p + M^*f(x)).$$

(3.5) <u>Lemma.</u> For $1 < p < \infty$ $\|Mf\|_p \leq E_p \left(\|f\|_p (2\pi)^{1/p} + \|M^*f\|_p\right)$

for f in $L^p(-\pi,\pi)$ where $E_p > 0$ is a constant independent of f.

Proof. This is an immediate consequence of (3.4) and Minkowski's inequality for integrals.

In chapter 4 we prove the following

(3.6) <u>Theorem</u>. Let $F \subset (-\pi,\pi)$ and let χ_F be the characteristic function of F. For every $y > 0$ and $1 < p < \infty$ we have

$$\lambda_{M^*\chi_F}(y) = m\{x \in (-\pi,\pi) \mid M^*\chi_F(x) > y\} \leq B_p^p\, y^{-p}\, (mF)$$

where $B_p > 0$ is a constant dependent on p but independent of F and y.

(3.7) <u>Lemma.</u> For $1 < p < \infty$ for each measurable set $E \subset (-\pi,\pi)$ we have $\|M\chi_E\|_p \leq F_p \|\chi_E\|_p$ where $F_p > 0$ is a constant independent of E.

<u>Proof</u>. This is an immediate consequence of (3.5), (3.6) and (1.9) with $p_o = q_o = ((p+1)/2)$, $p_1 = q_1 = (p+1)$ and $t = (1 - 1/p)$.

Fix integer $N > 0$.

Let $M_N f(x) = \max\limits_{o \leq n \leq N} |S_n(x;f)|$

Let α denote any simple function with domain $(-\pi,\pi)$ and range in $\{0,1. \ldots, N\}$. We say α is an Nth order simple function. Let

$T_\alpha f(x) = S_{\alpha(x)}(x;f) \quad x \in (-\pi,\pi)$.

Clearly, T_α is linear for every N^{th} order simple function.

(3.8) Lemma. For $1 < p < \infty$ $\|T_\alpha \chi_E\|_p \leq F_p \|\chi_E\|_p$ for each measurable set $E \subset (-\pi,\pi)$ and for each Nth order simple function α where F_p is that of (3.7).

Proof. This is immediate by (3.7) and the fact that $\|T_\alpha f\|_p \leq \|Mf\|_p$ for each $1 < p < \infty$, for each $f \in L_p(-\pi,\pi)$ and for each Nth order simple function α.

(3.9) Lemma. For $1 < p < \infty$ $\lambda_{T_\alpha \chi_E}(y) \leq F_p^p\, y^{-p} \|\chi_E\|_p^p$ for each measurable set $E \subset (-\pi,\pi)$ and for each Nth order simple function α where F_p is that of (3.8).

Proof. This is an immediate consequence of (3.8) and (1.7).

(3.10) Lemma. For $1 < p < \infty$ $\|T_\alpha f\|_p \leq C_p \|f\|_p$ for every simple function f in $L^p(-\pi,\pi)$ and for each Nth order simple function α where $C_p > 0$ depends only on p.

Proof. This is an immediate consequence of (3.9), (1.18) and (1.19) with $p_o = q_o = ((p+1)/2)$, $p_1 = q_1 = (p+1)$ and $t = (1 - 1/p)$.

(3.11) Lemma. Let $f \in L^p(-\pi,\pi)$ $1 < p < \infty$. Let α be any N^{th} order simple function. There exists a sequence of simple functions $\{f_n\} \subset L^p(-\pi,\pi)$ such that $\|f_n\|_p \to \|f\|_p$ and $T_\alpha(f-f_n)(x) \to 0$ for $x \in (-\pi,\pi)$.

Proof. This is an immediate consequence of the Lebesgue dominated convergence theorem and the fact that there exists a sequence of simple functions $\{f_n\} \subseteq L^p(-\pi,\pi)$ such that $\|f_n\|_p \to \|f\|_p$, $f_n(x) \to f(x)$ for $x \in (-\pi,\pi)$ and $|f_n(x)| \leq |f(x)|$ for $x \in (-\pi,\pi)$ and $n \geq 0$.

(3.12) <u>Lemma.</u> For $1 < p < \infty$ $\|T_\alpha f\|_p \leq C_p \|f\|_p$ for every f in $L^p(-\pi,\pi)$ and for each Nth order simple function α where C_p is that of (3.10).

<u>Proof.</u> Fix $f \in L^p(-\pi,\pi)$. Let $\{f_n\}$ be the sequence of (3.11). Then $|T_\alpha f_n(x)| \rightarrow |T_\alpha f(x)|$ for $x \in (-\pi,\pi)$; so that by Fatou's theorem and (3.10) we have

$$\|T_\alpha f\|_p \leq \lim_{n\to\infty} \|T_\alpha f_n\|_p \leq C_p \lim_{n\to\infty} \|f_n\|_p = C_p \|f\|_p$$

<u>Proof of Theorem (2.2)</u>

Fix f_o in $L^p(-\pi,\pi)$. It is easily shown that there exists an Nth order simple function α_o such that $|T_{\alpha_o} f_o(x)| = M_N f_o(x)$ for all $x \in (-\pi,\pi)$ Hence by (3.12) $\|M_N f_o\|_p \leq C_p \|f_o\|_p$. But $M_N f_o(x)$ increases monotonically to $Mf_o(x)$ for each x in $(-\pi,\pi)$. Hence $\|Mf_o\|_p \leq C_p \|f_o\|_p$.

IV. A PROOF OF THEOREM (3.6)

(4.1) Remark. The notation in this chapter is the same as that given at the beginning of chapter III.

In chapter 5 we will prove the following

(4.2) Theorem. Fix: $N > 0$, $1 < p < \infty$, $y > 0$ and $F \subset (-\pi,\pi)$. Let χ_F be the characteristic function of F. Then there exists a set $E \subset (-4\pi,4\pi)$ such that $m(E) \leq C_1^p y^{-p}(mF)$ and for all n such that $|n| \leq N$ and for all $x \in (-\pi,\pi) - E$ we have

$|S_n^*(x;\chi_F;\omega_{-1}^*)| \leq (C_2 L)y$ where $C_1 > 0$, $C_2 > 0$, $L > 0$ are

independent of N, y, and F, but $E = E(F, y, p, N)$ and $L = L(p)$.

Proof of Theorem (3.6).

Let $E_N = \{x \in (-\pi,\pi) \mid \sup\{|S_n^*(x;\chi_F;\omega_{-1}^*)| \, |n| \leq N\} > y\}$; $N > 0$

Let $E = \{x \in (-\pi,\pi) \mid M^*\chi_F(x) > y\}$.

It is easily shown that $E_{N+1} \supseteq E_N$ and $E = \bigcup_{N=1}^{\infty} E_N$; so that

since E is bounded, $m(E) = \lim_{N\to\infty} m(E_N)$. But by (4.2) we have for each $N > 0$

$$m\,E_N \leq (C_1 C_2 L)^p y^{-p}\ (mF)$$

Consequently, since (C_1C_2L) is independent of $N > 0$ we have that

$$\operatorname{Limit}_{N\to\infty} m\,E_N \leq (C_1C_2L)^p y^{-p}\ (mF).$$

V. A PROOF OF THEOREM (4.2)

(5.1) Notation. $|A|$ will denote the Lebesgue measure of $A \subset (-4\pi, 4\pi)$. For each integer $\nu \geq 0$ we subdivide $(-2\pi, 2\pi)$ into 2.2^{ν} equal intervals (called dyadic intervals) of length $2\pi \cdot 2^{-\nu}$. The resulting intervals are from left to right denoted $\omega_{j\nu}$, $j = 1, \ldots, 2.2^{\nu}$. Let $\omega^*_{-1} = (-4\pi, 4\pi)$ and for $j = 1, \ldots, (2.2^{\nu})-1$, $\nu \geq 0$ let $\omega^*_{j\nu} = \omega_{j\nu} \cup \omega_{j+1,\nu}$.

ω will always denote a dyadic interval $\omega_{j\nu}$ contained in $(-2\pi, 2\pi)$.

ω^* will always (except for ω^*_{-1}) denote the union of two adjacent dyadic intervals of common length. Consequently, for every ω^* there exists $\omega' \subset \omega^*$ such that for some $\nu \geq 0$ $|\omega'| = \left(\frac{2\pi}{2^{\nu+1}}\right)$ and $|\omega^*| = 2\left(\frac{2\pi}{2^{\nu}}\right)$;

or in other words for every ω^* (including ω^*_{-1}) there exists $\omega' \subset \omega^*$ such that $4|\omega'| = |\omega^*|$. Note that for ω^*_{-1} we have $[0, 2\pi] \subset \omega^*_{-1}$ and $4|[0, 2\pi]| = |\omega^*_{-1}|$. Also, for some $\nu \geq 0$ and for some $j \geq 1$ we have $\omega^*_{j\nu} \neq \omega_{\ell\mu}$ for all $\ell \geq 0$ and for all $\mu \geq 0$.

For each nonnegative integer n let $n[\omega_{j\nu}]$ be the greatest nonnegative integer less than or equal to $n2^{-\nu}$. Let

$n[\omega^*_{-1}] = n$. For $\nu \geq 0$ let $n[\omega^*_{j\nu}] = n[\omega_{1,\nu+1}]$.

Let $b_k = \dfrac{1}{2^k}$; $k = 0, 1, 2, \ldots$

For α real and $\omega = \omega_{j\nu}$ let

$$c_\alpha(\omega) = c_\alpha(\omega;f) = \frac{1}{|\omega|}\int_\omega f^o(x)\ \varepsilon^{-i2^\nu \alpha x}\,dx$$

For each pair $p = (n,\omega)$ we associate the number

$$C(p) = C_n(\omega) = C_n(\omega;f) = \frac{1}{10}\sum_{\mu=-\infty}^{\infty} |c_{(n+\frac{\mu}{3})}(\omega)|\ (1+\mu^2)^{-1}$$

Note that

$$0 \le C_n(\omega) \le \sup_{-\infty<\mu<\infty} |c_{(n+\frac{\mu}{3})}(\omega)| \le \left(\frac{1}{|\omega|}\int_\omega |f^o(x)|\ dx\right),$$ and

$C_n(\omega;f) = 0$ iff $f = 0$ almost everywhere in ω.

Let $C^*_n(\omega^*_{-1}) = C_n(\omega_{10}) = C_n(\omega_{20})$

For each pair $p^* = (n, \omega^*)$ we associate the number

$$C^*(p^*) = C^*_n(\omega^*) = \max\{C_n(\omega') \mid \omega' \subset \omega^*,\ 4|\omega'| = |\omega^*|\}$$

Note that $C^*_{n[\omega^*]}(\omega^*) = \max\{C_{n[\omega']}(\omega') \mid \omega' \subset \omega^*, 4|\omega'| = |\omega^*|\}$

By the statement: "x in the middle half of ω^*" we mean x is in one of the two middle fourths of ω^*.

Let $S^*_n(x;f;\omega^*) = \text{P.V.} \displaystyle\int_{\omega^*} \frac{\varepsilon^{-int} f^o(t)}{x-t}\,dt$; $|n| \ge 0$.

Note that $S^*_n(x;f;\omega^*)$ is the complex conjugate of

$S_{-n}(x;f;\omega^*)$ for $|n| \ge 0$.

In chapter 6 we will prove the following

(5.2) Theorem. Fix $N > 0$, $1 < p < \infty$, $y > 0$ and $F \subset (-\pi,\pi)$. Then there exists a set $E \subset (-4\pi,4\pi)$ such that $m(E) \le C_1^p\ y^{-p}(mF)$ and for each $x \in ((-\pi,\pi) - E)$ and for each n such that $0 \le n \le N$ there exists four finite sequences:

$$\omega^*_{-1},\ \omega^*_o,\ \omega^*_1,\ \ldots,\ \omega^*_j,\ \ldots,\ \omega^*_{J+1}$$

$$n = n_{-1},\ n_o,\ n_1,\ \ldots,\ n_j,\ \ldots,\ n_{J+1} = 0$$

$$k_{-1},\ k_o,\ k_1,\ldots,\ k_j,\ldots,\ k_{J+1}$$

$$m_{-1},\ m_o,\ m_1,\ \ldots,\ m_j,\ \ldots,\ m_{J+1}$$

(where the last three sequences consist of non negative integers.) such that

x is in the middle half of ω^*_j, $\omega^*_{j+1} \subset \omega^*_j$ (strictly)

$$k_{j+1} < m_j \le k_j \quad ; \quad n_j = 4\cdot 2\pi\cdot n_j[\omega^*_j]\ |\omega^*_j|^{-1} \quad ; \quad n_{j+1} \le (1 + b_{k_j})n_j$$

and $|S^*_{n_j}(x;\ \chi_F;\ \omega^*_j)| = |S^*_{n_{j+1}}(x;\ \chi_F;\ \omega^*_{j+1})| + O\ (Lm_j\, b_{m_{j-1}} y)$

for each j. In addition for each j

$|S^*_o(x;\chi_{F;}\ \omega^*_j)| = O\ (Ly)$ where $C_1 > 0$, $L > 0$ are independent of N, y, and F, but $E = E(F,y,p,N)$ and $L = L(p)$ and where the equation $r(x) = O\ (s(x))$ means $\left|\frac{r(x)}{s(x)}\right| \le C$ where $C > 0$ is independent of F, y, p, N and $x \varepsilon\ (-\pi,\pi)$.

<u>Proof of Theorem (4.2)</u>

<u>Case I.</u> $(n = 0)$

By (5.2) $|S^*_o(x;\ \chi_F;\ \omega^*_{-1})| = O\ (Ly)$

<u>Case II.</u> $(0 < n \le N)$

By (5.2) we have

$$|S^*_n(x;\ \chi_F;\ \omega^*_{-1})| = |S^*_o(x;\ \chi_F;\ \omega^*_{J+1})| + O\left(Ly \sum_{i=1}^{\infty} i\, b_{i-1}\right).$$

But $|S^*_o(x; \chi_F; \omega^*_{J+1})| = O\ (Ly)$ and $\sum_{i=1}^{\infty} i\, b_{i-1} < \infty$.

<u>Case III.</u> $(-N \le n < 0)$

This is immediate by Case II since

$$|S_n(x; \chi_F; \omega^*)| = |S_{-n}(x; \chi_F; \omega^*)|$$

(5.3) <u>Remark.</u> In general in (5.2) for each j we have $\omega^*_j \cap E \neq \phi$. In the proof of (4.2) Case II we take the entire infinite sum $\sum_{i=1}^{\infty} i\ b_{i-1}$ to insure the existence of an independent constant. Note that the condition $k_{j+1} < m_j \le k_j$ permits the use of the entire infinite sum. The conditions $n_j = 4 \cdot 2\pi \cdot n_j [\omega^*_j] |\omega^*_j|^{-1}$ and $n_{j+1} \le (1 + b_{k_j}) n_j$ play an important role in the algorithm needed to construct the four finite sequences in (5.2).

VI. A PROOF OF THEOREM (5.2)

The first six sections of this chapter are concerned with the construction of the exceptional set E in (5.2) and the estimate of its measure. We will construct this set as $E = (S^* \cup T^* \cup U^* \cup V^* \cup W^* \cup X^* \cup Y^*)$. Actually, the set $(T^* \cup U^* \cup V^*)$ is the "exceptional" set in the usual sense of the word (V^* is used only to establish the last statement in (5.2)). Roughly speaking, the set $(S^* \cup W^* \cup X^* \cup Y^*)$ is an "operational" set in the sense that if $x \in \omega^*$ but $x \notin E$, then $\omega^* \not\subset (S^* \cup X^* \cup Y^*)$ and $x \notin W^*$; and this fact will (among other things) allow certain "pair changing" operations to be performed. Consequently, the most important function of this "operational" set is its use in the proof of (6.40) and (6.41), the pair changing theorems.

In the rest of this monograph a lower case letter with a Greek subscript (for example: c_α) will denote a positive constant independent of F, y, p, N and $x \in (-\pi,\pi)$. Context will usually indicate when two lower case letters with the same subscript denote the same constant.

We fix $1 < p < \infty$, $y > 0$, $F \subset (-\pi,\pi)$ and $N > 0$. We assume $mF > 0$.

1. Construction of the $P_k(x;\omega)$ polynomials and the sets G_k, X_k^*

Fix integer $k \geq 1$.

Consider $\omega = \omega_{ro}$; $r=1,2$

Let $G_k(\omega_{ro}) = \{(n, \omega_{ro}) \mid |c_n(\omega_{ro}; \chi_F^o)| \geq b_k y^{p/2}\}$.

Note that since $\lim_{|n|\to\infty} |c_n(\omega_{ro}; \chi_F^o)| = o$, $G_k(\omega_{ro})$ is a finite set. Also, it is clear that $(n, \omega_{ro}) \in G_k(\omega_{ro})$ implies $(-n, \omega_{ro}) \in G_k(\omega_{ro})$ for $|n| \geq o$.

Let $P_k(x;\omega_{ro}) = \sum_{(n,\omega_{ro}) \in G_k(\omega_{ro})} c_n(\omega_{ro}; \chi_F^o)\, \varepsilon^{inx} = R_k(x;\omega_{ro})$; $|x| \leq 2\pi$

For $\omega_{s1} \subset \omega_{ro}$

Let $G_k(\omega_{s1}) = \{(n, \omega_{s1}) \mid |c_n(\omega_{s1}; \chi_F^o - P_k(\cdot\, ; \omega_{ro}))| \geq b_k y^{p/2}\}$.

Let $R_k(x; \omega_{s1}) = \sum_{(n,\omega_{s1}) \in G_k(\omega_{s1})} c_n(\omega_{s1}; \chi_F^o - P_k(\cdot\, ; \omega_{ro}))\, \varepsilon^{i2nx}$; $|x| \leq 2\pi$

Let $P_k(x;\omega_{s1}) = P_k(x;\omega_{ro}) + R_k(x;\omega_{s1}) = R_k(x;\omega_{ro}) + R_k(x;\omega_{s1})$.

For $\omega_{t2} \subset \omega_{s1}$

Let $G_k(\omega_{t2}) = \{(n,\omega_{t2}) \mid |c_n(\omega_{t2}; \chi_F^o - P_k(\cdot; \omega_{s1}))| \geq b_k y^{p/2}\}$.

Let $R_k(x;\omega_{t2}) = \sum_{(n,\omega_{t2}) \in G_k(\omega_{t2})} c_n(\omega_{t2}; \chi_F^o - P_k(\cdot;\omega_{s1}))\, \varepsilon^{i4nx}$; $|x| \leq 2\pi$

Let $P_k(x;\omega_{t2}) = P_k(x;\omega_{s1}) + R_k(x;\omega_{t2}) = R_k(x;\omega_{ro}) + R_k(x;\omega_{s1}) + R_k(x;\omega_{t2})$

Suppose $\omega_{jv} \subset \omega_{\ell v-1}$ and $P_k(x;\omega_{\ell v-1})$ has been constructed.

Let $G_k(\omega_{jv}) = \{(n,\omega_{jv}) \mid |c_n(\omega_{jv}; \chi_F^o - P_k(\cdot;\omega_{\ell,v-1}))| \geq b_k y^{p/2}\}$

Let $R_k(x;\omega_{jv}) = \sum_{(n,\omega_{jv})\varepsilon G_k(\omega_{jv})} c_n(\omega_{jv}, \chi_F^o - P_k(\cdot;\omega_{\ell v-1}))\, \varepsilon^{i2^v nx}; |x| \leq 2\pi$ ·

Let $P_k(x;\omega_{jv}) = P_k(x;\omega_{\ell,v-1}) + R_k(x;\omega_{jv})$ · Then

$$P_k(x;\omega_{jv}) = R_k(x;\omega_{ro}) + R_k(x;\omega_{s1}) + \ldots + R_k(x;\omega_{\ell,v-1}) + R_k(x;\omega_{jv})$$

Continuing in this way we define for each $k \geq 1$ and for each dyadic interval ω a polynomial $R_k(x;\omega)$, a polynomial $P_k(x;\omega)$ and a set $G_k(\omega)$.

(6.1) Remark. To simplify the notation we will often denote by $a_n(\omega)$ the coefficient of the $\varepsilon^{i2\pi|\omega|^{-1}nx}$ term in $R_k(x;\omega)$ corresponding to the element $(n,\omega)\ \varepsilon\ G_k(\omega)$; for simplicity we may at times write $a\,\varepsilon^{i\lambda x}$. It is to be noted that in the sequel whenever we write $a\,\varepsilon^{i\lambda x}$ as a term in $P_k(x;\omega)$ we always assume that the terms in this polynomial have not been "collected"; that is to say $a\,\varepsilon^{i\lambda x}$ is a term of $R_k(x;\omega')$ for some $\omega' \supset \omega$. With this convention in mind we have that if $a\,\varepsilon^{i\lambda x}$ is a term of $P_k(x;\omega)$ for some $\omega \subset [-2\pi, 2\pi]$, then $|a| \geq b_k\, y^{p/2}$. Also, it is easily shown that $|c_m(\omega; \chi_F^o - P_k(\cdot;\omega))| < b_k y^{p/2}$ for $|m| \geq 0$ and $\omega \subset [-2\pi, 2\pi]$

Let $G_k = \bigcup\{G_k(\omega) \mid \omega \subset [-2\pi, 2\pi]\}$·

(6.2) Theorem. $\displaystyle\sum_{(n,\omega)\varepsilon\, G_k} |a_n(\omega)|^2 |\omega| \leq 2mF.$

Proof. Suppose $\omega_{jv} \subset \omega_{\ell,v-1}$ where $v \geq o$ is arbitrary. Then

$$\int_{\omega_{jv}} |\chi^o_F(x)-P_k(x;\omega_{jv})|^2 dx = \int_{\omega_{jv}} |\chi^o_F(x)-P_k(x;\omega_{\ell,v-1}))-R_k(x;\omega_{jv})|^2 dx.$$

But it is easily shown that $\{\chi^o_F-P_k(\cdot;\omega_{\ell,v-1})-R_k(\cdot;\omega_{jv})\}$ and $R_k(\cdot;\omega_{jv})$ are orthogonal over ω_{jv}; so that by a straightforward expansion of the right side of the above equation we get

$$\int_{\omega_{jv}} |\chi^o_F(x)-P_k(x;\omega_{jv})|^2 dx = \int_{\omega_{jv}} |\chi^o_F(x)-P_k(x;\omega_{\ell,v-1})|^2 dx - \sum_{(n,\omega_{jv})\in G_k(\omega_{jv})} |a_n(\omega_{jv})|^2 |\omega_{jv}|$$

Hence

$$\sum_{|\omega|=2\pi 2^{-v}} \int_\omega |\chi^o_F(x)-P_k(x;\omega)|^2 dx = \sum_{|\omega|=2\pi 2^{-(v-1)}} \int_\omega |\chi^o_F(x)-P_k(x;\omega)|^2 dx - \sum_{\substack{(n,\omega)\in G_k(\omega)\\ |\omega|=2\pi 2^{-v}}} |a_n(\omega)|^2 |\omega| \quad ; \quad \text{for } v \geq 1.$$

We can now repeat the same argument for the first term on the right in the above equation. Finally, after a finite number of steps we have

$$0 \leq \sum_{|\omega|=2\pi 2^{-v}} \int_\omega |\chi^o_F(x)-P_k(x;\omega)|^2 dx = \int_{-2\pi}^{2\pi} |\chi^o_F(x)|^2 dx - \sum_{\substack{(n,\omega)\in G_k\\ |\omega|\geq 2\pi 2^{-v}}} |a_n(\omega)|^2 |\omega| .$$

But since $v \geq o$ is arbitrary, the result follows.

(6.3) Corollary. $\sum_{(n,\omega)\varepsilon\, G_k} |\omega| \leq 2\, b_k^{-2} y^{-p} mF.$

Proof. Since $|a_n(\omega)| \geq b_k\, y^{p/2}$ if $(n,\omega)\ \varepsilon\ G_k$ we have

$$|a_n(\omega)|^2 \geq b_k^2\, y^p \ ; \quad \text{so that } 1 \leq b_k^{-2} y^{-p}\ |a_n(\omega)|^2 \ ;$$

so that $|\omega| \leq b_k^{-2} y^{-p}\ |a_n(\omega)|^2 |\omega|$. Consequently, by (6.2)

$$\sum_{(n,\omega)\ \varepsilon\ G_k} |\omega| \leq b_k^{-2} y^{-p} \sum_{(n,\omega)\ \varepsilon\ G_k} |a_n(\omega)|^2 |\omega| \leq 2\ b_k^{-2} y^p\ mF.$$

This completes the proof of (6.3).

For $x\ \varepsilon\ [-2\pi, 2\pi]$ and for $v \geq 0$ we define

$$A_k^v(x) = \begin{cases} \sum_{(n,\omega)\ \varepsilon\ G_k(\omega)} |a_n(\omega)|^2 & \text{if } x\ \varepsilon\ \omega^o \text{ and } |\omega| = 2\pi 2^{-v} \\ 0 & \text{if } x \text{ is an end point of } \omega \text{ and } |\omega| = 2\pi 2^{-v} \end{cases}$$

Note that for each $v \geq 0$ $A_k^v(x)$ is a simple function.

Let $A_k(x) = \sum_{v=o}^{\infty} A_k^v(x)$.

Let $X_k = \{x \mid A_k(x) > b_k^{-1}\, y^p\}$.

(6.4) Lemma. $m X_k \leq 2\, b_k\, y^{-p}\ mF.$

Proof.

$$\int_{-2\pi}^{2\pi} A_k(x)dx = \sum_{v=o}^{\infty} \int_{-2\pi}^{2\pi} A_k^v(x)dx = \sum_{v=o}^{\infty} \left(\sum_{\substack{(n,\omega)\ \varepsilon\ G_k(\omega) \\ |\omega|=2\pi 2^{-v}}} |a_n(\omega)|^2 |\omega| \right)$$

$$= \sum_{(n,\omega)\ \varepsilon\ G_k} |a_n(\omega)|^2 |\omega|$$

Consequently, by (6.2) we have

$$\int_{-2\pi}^{2\pi} A_k(x)dx \leq 2\ m\ F$$

But if $x \in X_k$, we have $b_k y^{-p} A_k(x) > 1$; so that

$$m\ X_k \leq \int_{X_k} b_k y^{-p} A_k(x)dx \leq \int_{-2\pi}^{2\pi} b_k y^{-p} A_k(x)dx \leq 2b_k y^{-p} mF.$$

(6.5) Lemma. If $\omega \notin X_k$, then $P_k(x;\omega)$ has at most b_k^{-3} terms.

<u>Proof.</u> Let $P_k(x;\omega) = \sum_{j=1}^{J} a_j \varepsilon^{i\lambda_j x}$ By construction for $1 \leq j \leq J$, $|a_j| \geq b_k y^{p/2}$; so that $|a_j|^2 b_k^{-2} y^{-p} > 1$; consequently, $J \leq \sum_{j=1}^{J} |a_j|^2 b_k^{-2} y^{-p} = b_k^{-2} y^{-p} \sum_{j=1}^{J} |a_j|^2$ Let $x_o \varepsilon\ (\omega - X_k)$. Then $A_k(x_o) \leq b_k^{-1} y^p$. But since $x_o \varepsilon\ \omega$ we have $\sum_{j=1}^{J} |a_j|^2 \leq A_k(x_o)$. Consequently,

$$J \leq b_k^{-2} y^{-p} A_k(x_o) \leq (b_k^{-2} y^{-p})(b_k^{-1} y^p) = b_k^{-3}$$

(6.6) <u>Lemma.</u> If $\omega \notin X_k$, then $|P_k(x;\omega)| \leq b_k^{-2} y^{p/2}$ for $|x| \leq 2\pi$.

<u>Proof.</u> Let $P_k(x;\omega) = \sum_{j=1}^{J} a_j \varepsilon^{i\lambda_j x}$. Then $|P_k(x;\omega)| \leq \sum_{j=1}^{J} |a_j|$ for $|x| \leq 2\pi$. By construction $|a_j| \geq b_k y^{p/2}$; so that $|a_j|\, b_k^{-1} y^{-p/2} > 1$ for $1 \leq j \leq J$. Consequently, $|P_k(x;\omega)| \leq b_k^{-1} y^{-p/2} \sum_{j=1}^{J} |a_j|^2$. Let $x_o \varepsilon\ (\omega - X_k)$. Then $\sum_{j=1}^{J} |a_j|^2 \leq A_k(x_o)$. Consequently, since

$$A_k(x_o) \le b_k^{-1}y^p,\ |P_k(x;\omega)| \le b_k^{-1}y^{-p/2}A_k(x_o) \le (b_k^{-1}y^{-p/2})(b_k^{-1}y^p) = b_k^{-2}y^{p/2}$$

(6.7) Remark. It is immediate by the definition of $A_k(x)$ that if $x \in X_k$, then there exists a dyadic interval $\omega \subset X_k$ with $x \in \omega$. For each $\omega \subset X_k$ we consider its three left dyadic neighbors $\omega_\ell^1, \omega_\ell^2, \omega_\ell^3$ and its three right dyadic neighbors $\omega_r^1, \omega_r^2, \omega_r^3$ all of length $|\omega|$.

Let $X_\omega = \omega_r^1 \cup \omega_r^2 \cup \omega_r^3 \cup \omega \cup \omega_\ell^1 \cup \omega_\ell^2 \cup \omega_\ell^3$.

If ω is located too close to either 2π or -2π, then some or all of the three left or right dyadic neighbors may not exist. If this situation occurs, simply delete the missing terms from the expression for $X\omega$.

It is clear that $|X_\omega| \le 7|\omega|$.

Let $X_k^* = \bigcup\{X_\omega | \omega \subset X_k\}$.

(6.8) Lemma. $m X_k^* \le 14\, b_k y^{-p}\, mF$.

Proof. $m\, X_k^* \le 7\, m\, X_k \le 7\,(2\, b_k y^{-p} mF)$ by (6.4).

(6.9) Remark. Note that if $\omega^* \not\subset X_k^*$, then $\omega' \not\subset X_k$ for each of the four subintervals ω' of ω^* with $4|\omega'| = |\omega^*|$.

2. Construction of the sets G_k^* and Y_k^*.

(6.10) Remark. We first note that if $P_k(x;\omega)$ contains a term $a\,\varepsilon^{i\lambda x}$, then it also contains the term $\bar{a}\,\varepsilon^{i(-\lambda)x}$.

Also, if $\lambda > 0$, then $(\lambda[\omega'], \omega') \in G_k$ for some $\omega \subset \omega'$; for by construction $\lambda = 2^v n$ for some integer $v > 0$ and for some integer $n > 0$ and $\omega_{jv} \supset \omega$ and $(n, \omega_{jv}) \in G_k(\omega_{jv})$. But $(2^v n)[\omega_{jv}] = n$. Hence $(\lambda[\omega_{jv}], \omega_{jv}) \in G_k(\omega_{jv})$. On the other hand for each $(n', \omega') \in G_k$ where $n' > 0$ let $\lambda = n' 2\pi |\omega'|^{-1}$. Then $n' = \lambda[\omega']$.

For each $k \geq 1$ consider the following two conditions on a pair $p = (n, \omega)$:

$$(A_k) \begin{cases} \text{For some } (\lambda[\omega'], \omega') \in G_k : \omega \subset \omega',\ n > 0,\ |n - \lambda[\omega]| < b_k^{-10} \\ \text{and } |\omega| > b_k^{10} |\omega'|. \end{cases}$$

$$(B_k) \begin{cases} \text{For some } (\lambda[\omega'], \omega') \in G_k : \omega \subset \omega',\ n > 0,\ |n - \lambda[\omega]| < b_k^{-10} \\ \text{and there is some term } a' \, \varepsilon^{i\lambda' x} \text{ of } P_k(x; \omega') \text{ such that} \\ b_k^{10} \leq |\lambda - \lambda'| |\omega| \leq b_k^{-20}. \end{cases}$$

We let $\tilde{G}_k = \{(n, \omega) \mid \omega \not\subset X_k$ and (n, ω) satisfies "A_k" or "B_k"$\}$

(6.11) <u>Lemma.</u> $\displaystyle\sum_{(n,\omega) \in \tilde{G}_k} |\omega| \leq C_\theta \, b_k^{-19} \, y^{-p} \, mF.$

For each $k \geq 1$ let

$$G_k^* = \{(n, \omega^*) \mid (n, \omega') \in \tilde{G}_k, \omega^* \supset \omega' \text{ and } 4|\omega'| = |\omega^*|\}$$

(6.12) <u>Remark.</u> If ω' is not located too close to either 2π or -2π, then there exists two intervals ω^* such that $\omega^* \supset \omega'$ and $4|\omega'| = |\omega^*|$. Also, if $\omega^* \not\subset X_k^*$ and $(n, \omega^*) \notin G_k^*$, then $\omega' \not\subset X_k$ and $(n, \omega') \notin \tilde{G}_k$ for each of the four intervals $\omega' \subset \omega^*$, $4|\omega'| = |\omega^*|$. Note that $(n,\omega^*_{-1}) \in G^*_k$ iff $(n,\omega_{10}) \in \tilde{G}_k$ and $(n,\omega_{20}) \in \tilde{G}_k$.

(6.13) Lemma. $\sum_{(n,\omega^*) \in G^*_k} |\omega^*| \leq c_\delta\, b_k^{-19}\, y^{-p}\, (mF)$

Proof. This is immediate by (6.11).

Let ω be any dyadic interval contained in $[-2\pi, 2\pi]$.

Let F^1_ω be the interval of length $2\, b_k^3 |\omega|$ symmetric about the left end point of ω.

Let F^2_ω be the interval of length $2\, b_k^3 |\omega|$ symmetric about the right end point of ω.

Let $F_\omega = F^1_\omega \cup F^2_\omega$

Let $Y^*_k = \bigcup \{F_\omega \mid (n,\omega) \in G_k$ for some integer $n\}$

(6.14) Lemma. $m\, Y^*_k \leq 8\, b_k y^{-p}\, mF$.

Proof. Clearly, $m\, Y^*_k \leq 4 b_k^3 \sum_{(n,\omega) \in G_k} |\omega|$. But by (6.3) we have $\sum_{(n,\omega) \in G_k} |\omega| \leq 2\, b_k^{-2} y^{-p} mF$. Hence $mY^*_k \leq (4b_k^3)\, (2b_k^{-2} y^{-p}\, mF)$.

2. Construction of the sets S^* and $\Omega(k)$.

Let $S = \bigcup \{\omega \mid y^{-p} \int_\omega |\chi^o_F(x)|\, dx \geq |\omega| \}$.

(6.15) Lemma. $m\, S \leq 2\, y^{-p} mF$

Proof. $m\, S \leq y^{-p} \int_{-2\pi}^{2\pi} |\chi^o_F(x)|\, dx = 2\, y^{-p}\, mF$

(6.16) Remark. For each $\omega \subset S$ we consider its three left dyadic neighbors $\omega^1_\ell, \omega^2_\ell, \omega^3_\ell$ and its three right dyadic neighbors $\omega_r^1, \omega_r^2, \omega_r^3$ all of length $|\omega|$.

Let $S_\omega = \omega_r^1 \cup \omega_r^2 \cup \omega_r^3 \cup \omega \cup \omega_\ell^1 \cup \omega_\ell^2 \cup \omega_\ell^3$

If ω is located too close to either 2π or -2π, then some or all of the three left or right dyadic neighbors may not exist. If this situation occurs, simply delete the missing terms from the expression for S_ω.

It is clear that $|S_\omega| \le 7|\omega|$.

Let $S^* = \bigcup \{S_\omega | \omega \subset S\}$.

(6.17) Lemma. $mS^* \le 14y^{-p} mF$.

Proof. $mS^* \le 7mS \le 7(2y^{-p}mF)$ by (6.15).

(6.18) Lemma. If $\omega \not\subset S$, then $C(\omega\ ; \chi_F^o) < y$

Proof. By definition of S we have that

$$\left(\frac{1}{|\omega|}\right)^{1/p} \|\chi_F^o\|_p < y$$

and by Holder's inequality it is easily shown that

$$\|\chi_F^o\|_1 \le \left(\frac{1}{|\omega|}\right)^{(1/p-1)} \|\chi_F^o\|_p$$

But since $C(\omega\ ; \chi_F^o) \le \sup_{-\infty<\mu<\infty} | c_{n+\frac{\mu}{3}}(\omega)| \le \frac{1}{|\omega|} \|\chi_F^o\|_1$

we have

$$C(\omega\ ; \chi_F^o) \le \frac{1}{|\omega|} \|\chi_F^o\|_1$$

Hence

$$C(\omega; \chi_F^o) \le \frac{1}{|\omega|}\left(\frac{1}{|\omega|}\right)^{(1/p-1)} \|\chi_F^o\|_p = \left(\frac{1}{|\omega|}\right)^{1/p} \|\chi_F^o\|_p < y\,.$$

(6.19) <u>Lemma</u>. If $\omega^* \not\subset S^*$ and $\chi_F^o \not\equiv 0$ a.e. on ω^*, then for some $k \ge 1$ we have

$$b_k y \le C_n^*(\omega^*; \chi_F^o) < b_{k-1}\, y$$

<u>Proof.</u> Since $\chi_F^o \not\equiv 0$ a.e. on ω^*, we know $0 < C_n^*(\omega^*; \chi_F^o)$.

But since $\omega^* \not\subset S^*$ we have $\omega' \not\subset S$ for all four $\omega' \subset \omega^*$, $4|\omega'| = |\omega^*|$. Hence by (6.18) we have $0 < C_n(\omega'; \chi_F^o) < y$ for all four $\omega' \subset \omega^*$, $4|\omega'| = |\omega^*|$. Consequently, $0 < C_n^*(\omega^*; \chi_F^o) < y$. The lemma now follows immediately from the definition of b_k.

(6.20) <u>Lemma.</u> There exists a positive integer L which is a function p only such that if $\omega^* \not\subset S^*$, then

$$b_k y \le C_n^*(\omega^*; \chi_F^o) \text{ implies } y^{p/2} \le b_{kL}^{-1/4}\, y.$$

We now define for each $k \ge 1$ the set

$$\Omega(k) = \{p^* = (n[\omega^*], \omega^*) \in G_{kL}^* \mid C^*(p^*) < b_{k-1} y \text{ and } n = 4\cdot 2\pi.\ |\omega^*|^{-1} \cdot n[\omega^*]\}$$

4. <u>Construction of the partition $\Omega(p^*,k)$ of ω^* for each p^* in $\Omega(k)$.</u>

Fix $k \ge 1$. Let $p^* = (n[\omega^*], \omega^*) \in \Omega(k)$. To construct the partition $\Omega(p^*, k)$ of ω^* we require that each interval ω' of our partition satisfy

(6.21) $$C_{n[\omega']}(\omega') < b_{k-1}\, y.$$

Clearly, each of the four intervals $\omega' \subset \omega^*$, $4|\omega'| = |\omega^*|$ satisfy (6.21). For each of these intervals ω' we consider the two intervals $\omega'' \subset \omega'$, $2|\omega''| = |\omega'|$. If each of these two intervals satisfy (6.21), we split ω'; otherwise ω' is an interval of our partition. We continue splitting according to the above rule as long as possible or until we reach an interval of length $2\pi\, 2^{-N}$. In addition to (6.21) each interval ω' of our partition will satisfy

(6.22) $\quad |\omega'| \geq 2\pi\, 2^{-N}$

(6.23) $\left\{\begin{array}{l} \text{If } \frac{1}{4}|\omega^*| \geq |\omega'| \geq 2\pi\, 2^{-N+1}, \text{ then (6.21) does not} \\ \text{hold for at least one of the two intervals } \omega'' \subset \omega', \\ 2|\omega''| = |\omega|, \text{ and} \end{array}\right.$

(6.24) If $\omega^* \supset \tilde{\omega} \supset \omega'$ and $4|\tilde{\omega}| \leq |\tilde{\omega}^*|$, then (6.21) holds for $\tilde{\omega}$.

We now define $\omega^*(x)$, <u>the center interval corresponding to $\Omega(p^*,k)$ and x</u>

Consider the collection of intervals $\tilde{\omega}^*$ which are formed by taking each $\omega_{j\nu} \in \Omega(p^*,k)$ and adjoining $\omega_{j-1,\nu}$ or $\omega_{j+1,\nu}$. For each x in the middle half of ω^*, there are intervals (at least one) $\tilde{\omega}^*$ as above, which contain x in their middle half. We define $\omega^*(x)$ as such an interval $\tilde{\omega}^*$ with $|\tilde{\omega}^*|$ maximal. We have

(6.25) $\quad 2|\omega^*(x)| \leq |\omega^*|$

(6.26) x belongs to the middle half of $\omega^*(x)$.

(6.27) $\left\{ \begin{array}{l} \omega^*(x) \text{ is a union of intervals of } \Omega(p^*,k), \text{ since} \\ |\omega^*(x)| \text{ is maximal.} \end{array} \right.$

(6.28) $\left\{ \begin{array}{l} \text{If } \omega^*(x) = \omega_{jv} \cup \omega_{j\pm1,v}, \text{ where } \omega_{jv} \in \Omega(p^*,k), \\ \text{it follows from (6.21), (6.24), and (6.27) that} \\ \max \left\{ C_{n[\omega_{jv}]}(\omega_{jv}),\ C_{n[\omega_{j+1,v}]}(\omega_{j\pm1,v}) \right\} < b_{k-1}\, y. \end{array} \right.$

(6.29) $\left\{ \begin{array}{l} \text{If } |\omega^*(x)| > 2\cdot 2\pi \cdot 2^{-N} \text{ it follows from (6.23) that} \\ C^*_{n[\omega^*(x)]}(\omega^*(x)) > b_{k-1} y\ . \end{array} \right.$

$\omega^* - \omega^*(x)$ is by (6.27) the union of certain intervals of $\Omega(p^*,k)$.

(6.30) $\left\{ \begin{array}{l} \text{For each such interval } \omega' \text{ the distance from } x \text{ to } \omega' \\ \text{exceeds half the length of } \omega', \text{ since } |\omega^*(x)| \text{ is maximal.} \end{array} \right.$

5. <u>Construction of the sets T^* and U^*.</u>

Suppose $p^* = (n[\omega^*], \omega^*) \in \Omega(k)$ for some $k \geq 1$.

Let $\Omega(p^*,k)$ be the partition of ω^*. For $t \in \omega^*$ we let

$$E_n(t) = \frac{1}{|\omega_m|} \int_{\omega_m} \chi^{\,o}_{F}(y)\ \varepsilon^{-iny}\,dy \quad t \in \omega_m \in \Omega(p^*,k)$$

(6.31) <u>Lemma</u> $|E_n(t)| \leq C_\theta\, b_{k-1} y$ for $t \in \omega^*$.

For each $\omega_m \in \Omega(p^*,k)$ let ω_m have midpoint t_m and let $|\omega_m| = \delta_m$.

For $x \in \omega^*$ let $\Delta(x) = \Delta(x;\ \Omega) = \sum_m \dfrac{\delta_m^2}{(x-t_m)^2 + \delta_m^2}$

Let C be a fixed positive constant (to be specified in the sequel).

Let $T^*(p^*) = \{x \in \omega^* \mid H_n^*(x) > CL\,k\,b_{k-1}\,y\}$ where $H_n^*(x)$ is

the maximal Hilbert transform of $E_n(t)$ over ω^*.

Let $U^*(p^*) = \{x \in \omega^* \mid \Delta(x) > C\,L\,k\}$

(6.32) Lemma. $mT^*(p^*) \leq c_\alpha \, \varepsilon^{-c_\beta CLk} |\omega^*|$ and

$mU^*(p^*) \leq c_\alpha \, \varepsilon^{-c_\beta CLk} |\omega^*|$.

(6.33) Theorem. If $x \in (-\pi,\pi)$ and $x \notin (T^*(p^*) \cup U^*(p^*))$ then

$$\big|\,|S_n^*(x;\chi_F;\omega_o^*)| - |S_n^*(x;\chi_F;\omega^*(x)|\,\big| \leq C_\phi (L\,k\,b_{k-1}y)$$

where ω_o^* is any interval which satisfies

x is the middle half of ω_o^*, $\omega_o^* \subset \omega^*$, $\omega^*(x) \subset \omega_o^*$ (strictly)

and $\omega_o^* - \omega^*(x)$ is a union of intervals of $\Omega(p^*,k)$.

We let $T^* = \bigcup_{k=1}^{\infty} \big(\bigcup_{\Omega(k)} T^*(p^*)\big)$ and $U^* = \bigcup_{k=1}^{\infty} \big(\bigcup_{\Omega(k)} U^*(p^*)\big)$

(6.34) Lemma. $m(T^* \cup U^*) \leq c\,\Delta\, y^{-p} mF$.

Proof. Since $\Omega(k) \subset G^*_{kL}$ for each $k \geq 1$ we have by (6.13)

$$\Sigma|\omega^*| \leq \sum_{(n,\omega^*)\in G^*_{kL}} |\omega^*| \leq c_\delta \, b_{kL}^{-19} \, y^{-p} mF$$

$(n[\omega^*], \omega^*) \in \Omega(k)$

Hence by (6.32)

$$m\Big(\bigcup_{\Omega(k)} (T^*(p^*) \cup U^*(p^*))\Big) \leq (c'_\alpha \, \varepsilon^{-c_\beta CLk}) \sum_{(n[\omega^*],\omega^*)\in\Omega(k)} |\omega^*|$$

$$\leq (c'_\alpha \varepsilon^{-c_\beta CLk})(c_\delta \, b_{kL}^{-19} \, y^{-p} mF\,).$$

We now choose C such that $c_\beta C \geq 20 \log_\varepsilon 2$; so that

$$\left(\varepsilon^{-c_\beta CLk} \cdot b_{kL}^{-19}\right) \leq b_{kL}$$

Hence $m(T^* \cup U^*) \leq c'_\Delta \left(\sum_{k=1}^{\infty} b_{kL}\right) y^{-p} mF \leq c_\Delta y^{-p} mF.$

6. Construction of the exceptional set E.

We let $X^* = \bigcup_{k=1}^{\infty} X^*_{kL}$ and $Y^* = \bigcup_{k=1}^{\infty} Y^*_{kL}$

(6.35) Lemma. $m(X^* \cup Y^*) \leq c_\Sigma y^{-p} mF$.

Proof. By (6.8) and (6.14) we have

$$m(X^* \cup Y^*) \leq c'_\Sigma \left(\sum_{k=1}^{\infty} b_{kL}\right) y^{-p} mF \leq c_\Sigma y^{-p} mF.$$

Let $V^* = \{x \in \omega^*_{-1} \mid H^* \chi^o_F(x) > Ly\}$ where $H^* \chi^o_F(x)$ is the maximal Hilbert transform of $\chi^o_F(t)$ over ω^*_{-1}.

(6.36) Lemma. $mV^* \leq K^p_p \; y^{-p} mF$

Let $W^* = \{x \mid x$ is an endpoint of some dyadic interval $\omega\}$

(6.37) Lemma. $m(W^*) = 0$.

Proof. This is immediate since W^* is countable.

Let $E = S^* \cup T^* \cup U^* \cup V^* \cup W^* \cup X^* \cup Y^*$.

(6.38) Theorem. $m(E) \leq C_p \; y^{-p} \; mF$.

Proof. This is immediate by (6.17), (6.34), (6.35), (6.36), (6.37).

7. The pair changing theorems.

Suppose n_o, ω^*_o, k and x satisfy

$$(6.39)\quad \begin{cases} x \notin E,\ x \text{ is in the middle half of } \omega_o^*, n_o = 4\cdot 2\pi\cdot|\omega_o^*|^{-1} n_o[\omega_o^*], \\ |\omega_o^*| > 2\cdot 2\pi\cdot 2^{-N},\quad p_o^* = (n_o[\omega_o^*], \omega_o^*) \notin G_{kL}^* \text{ and} \\ b_k y \le C^*(p_o^*) < b_{k-1} y. \end{cases}$$

(6.40) Theorem. Suppose n_o, ω_o^*, k and x satisfy (6.39). Then there exist $\tilde{n}$ and $\tilde{\omega}^*$ such that x is in the middle half of $\tilde{\omega}^*$, $\tilde{\omega}^* \supset \omega_o^*$, $\tilde{n} = 4\cdot 2\pi\cdot|\tilde{\omega}^*|^{-1}\tilde{n}[\tilde{\omega}^*]$, $\tilde{p}^* = (\tilde{n}[\tilde{\omega}^*], \tilde{\omega}^*) \in G_{kL}^*$

and $|\tilde{n}[\omega_o^*] - n_o[\omega_o^*]| \le A\, b_k^{-1}$, where A is a fixed integer. Moreover, if $\tilde{p}_o^* = (\tilde{n}[\omega_o^*], \omega_o^*)$, then

$$\big|\,|S_{n_o}^*(x;\omega_o^*)| - |S_n^*(x;\omega_o^*)|\,\big| \le c_\mu \{C^*(\tilde{p}_o^*) + b_{k-1} y\}$$

for all n such that $|n_o[\omega_o^*] - n[\omega_o^*]| \le 2\, A\, b_k^{-2}$.

(6.41) Theorem. Suppose n_o, ω_o^*, k and x satisfy (6.39). Then there exist $\bar{n}, \bar{\omega}^*$ and m such that x is in the middle half of $\bar{\omega}^*, \bar{\omega}^* \supset \omega_o^*$

$\bar{n} = 4\cdot 2\pi\cdot|\bar{\omega}^*|^{-1}\bar{n}[\bar{\omega}^*]$, $|\bar{n}[\omega_o^*] - n_o[\omega_o^*]| < 2A\, b_k^{-1}$, $\bar{p}^* = (\bar{n}[\bar{\omega}^*], \bar{\omega}^*) \in G_{mL}^*$, and $1 \le m \le k$. If $\tilde{p}_o^*$ is given by (6.40) then $C^*(\tilde{p}_o^*) < b_{m-1} y$.

Moreover, $C^*(\bar{p}^*) < b_{m-1} y$, so the partition $\Omega(\bar{p}^*; m)$ is defined. For this partition we have $\bar{\omega}^*(x) \subset \omega_o^*$ (strictly) and $\omega_o^* - \bar{\omega}^*(x)$ is a union of intervals of $\Omega(\bar{p}^*; m)$.

7. A Proof of Theorem (5.2).

(6.42) Lemma. If $\omega = \omega_{jv}$, then for each $n \ge 0$ we have $C_n(\omega) \le c_\alpha\, C_{n+1}(\omega)$ where c_α is also independent of ω.

(6.43) Lemma. Suppose $\omega^* = \omega_{jv} \cup \omega_{j+1,\,v}$ and $n_o[\omega^*] = \dfrac{n_o}{2^{v+1}}$.

Then for all n such that $|n-n_o| < 2^{v+1}$ and for $x \in (-\pi,\pi)$ we have

$$\left| |S_n^*(x;\chi_F;\omega^*)| - |S_{n_o}(x;\chi_F;\omega^*)| \right| \leq$$

$$c_\beta \left(\max \left\{ C_{n_o[\omega_{jv}]}(\omega_{jv}),\ C_{n_o[\omega_{j+1,v}]}(\omega_{j+1,v}) \right\} \right)$$

(6.44) Remark. Suppose for some $n > 0$ we have $\dfrac{n_o}{2^{v+1}} = n[\omega_o^*]$

where $\omega_o^* = (\omega_{jv} \cup \omega_{j+1,v})$. Then it is easily shown that $0 < (n-n_o) < 2^{v+1}$, $n_o[\omega_o^*] = n[\omega_o^*]$, and $n[\omega_{jv}] = n_o[\omega_{jv}]$ or

$n[\omega_{jv}] = n_o[\omega_{jv}] + 1$, and $n[\omega_{j+1,v}] = n_o[\omega_{j+1,v}]$ or $n[\omega_{j+1,v}] = n_o[\omega_{j+v,v}] + 1$. For example, if we let

$n_o = 4\cdot 2\pi\cdot |\omega_o^*|^{-1}\, n[\omega_o^*]$ where $\omega_o^* = (\omega_{jv} \cup \omega_{j+1,v})$, then

it is easily seen that $\dfrac{n_o}{2^{v+1}} = n[\omega_o^*]$.

(6.45) Remark. Note that the condition in (5.2) that

$n_j = 4\cdot 2\pi\cdot n_j[\omega_j^*]|\omega_j^*|^{-1}$ implies that $n_j[\omega_j^*] = \dfrac{n_j}{2^{v+1}}$ for

some $v \geq 0$ where $\omega_j^* = (\omega_{jv} \cup \omega_{j+1,v})$ and that $n_j = 0$ if and only if $n_j[\omega_j^*] = 0$.

(6.46) Remark. Using the fact that $\prod_{i=1}^{\infty}(1 + (\frac{1}{2})^i) < 2$

the condition in (5.2) that $n_{j+1} \leq (1 + b_{k_j})n_j$ implies

(since $n_{-1} = n$) that $n_j \leq \prod_{i=1}^{\infty} (1 + b_i)n \leq 2N \leq 2^N$. Consequently,

if $|\omega_j^*| \leq 2\cdot 2\pi\cdot 2^{-N}$, then by combining this with the condition in

(5.2) that $n_j = 4\cdot 2\pi\cdot n_j[\omega_j^*]|\omega_j^*|^{-1}$ we have that

$n_j[\omega_j^*]2^{N+1} \le 4\cdot 2\pi\ |\omega_j^*|^{-1}\ n_j[\omega_j^*] \le 2^N$. Hence $n_j[\omega_j^*] \le \frac{1}{2}$; so that $n_j[\omega_j^*] = 0$; so that $n_j = 0$. An equivalent statement is $n_j \neq 0$ implies $|\omega_j^*| > 2\cdot 2\pi\cdot 2^{-N}$. We now prove (5.2) by means of the following algorithm: Let $n_{-1} = n$ and $\omega^*_{-1} = [-4\pi, 4\pi]$. By (6.19) there exists k such that

$$b_k y \le C^*_n(\omega^*_{-1}) < b_{k-1} y.$$

(6.47) <u>Lemma.</u> $(n[\omega^*_{-1}], \omega^*_{-1}) \in G^*_{kL}$.

By (6.47) the partition $\Omega((n[\omega^*_{-1}], \omega^*_{-1}); k)$ is defined; so that by (6.33) since $x \notin E$ we have

$$|S^*_{n_{-1}}(x; \chi_F; \omega^*_{-1})| = |S^*_{n_{-1}}(x; \chi_F; \omega^*(x))| + O(L\ k\ b_{k-1} y)$$

We let $k_{-1} = m_{-1} = k, \omega^*(x) = \omega^*_o$ and $n_o = 4\cdot 2\pi\cdot n[\omega^*_o]|\omega^*_o|^{-1}$.

By (6.28), (6.42), (6.43) and (6.44) we have

$$|S^*_{n_{-1}}(x; \chi_F : \omega^*_{-1})| = |S^*_{n_o}(x; \chi_F; \omega^*_o)| + O\ (Lk\ b_{k-1} y)\ .$$

Suppose $n_o \neq 0$. Then by (6.46) we have $|\omega^*_o| > 2\cdot 2\pi\cdot 2^{-N}$; Consequently, by (6.29) we have $C_{n[\omega^*_o]}(\omega^*_o) \ge b_{k-1} y$. But this implies that $\chi^o_F \neq 0$ a.e. on ω^*_o ; so that by (6.19) there exists k_o such that $b_{k_o} y \le C^*_{n_o[\omega^*_o]}(\omega^*_o) < b_{k_o-1} y$.

Clearly, $k_o < k = m_{-1}$. Let $p^*_o = (n[\omega^*_o], \omega^*_o)$.

There are now three possibilities:

Case I. $p_o^* \varepsilon G_{k_oL}^*$

Then the partition $\Omega(p_o^*, k_o)$ is defined; so that by (6.33) we have $|S_{n_o}^*(x; \chi_F; \omega_o^*)| = |S_{n_o}^*(x; \chi_F; \omega_o^*(x))| + O(L k_o b_{k_o-1}y)$.

Let $\omega_1^* = \omega_o^*(x)$, $m_o = k_o$, and $n_1 = 4\cdot 2\pi\cdot n_o[\omega_1^*] |\omega_1^*|^{-1}$.

By (6.28, (6.42), (6.43), and (6.44) we have

$$|S_{n_o}^*(x; \chi_F; \omega_o^*)| = |S_{n_1}^*(x; \chi_F; \omega_1^*)| + O(L m_o b_{m_o-1}y) .$$

Case 2. $p_o^* \notin G_{k_oL}^*$ and $n_o[\omega_o^*] > 2 A b_{k_o}^{-2}$.

Choose $\tilde{n}$ as in (6.40) and $\bar{n}$, $\bar{\omega}^*$, m as in (6.41). The partition $\Omega(\bar{p}^*; m)$ yields by (6.33)

$$|S_{\bar{n}}^*(x; \chi_F; \omega_o^*)| = |S_{\bar{n}}^*(x; \chi_F; \bar{\omega}^*(x))| + O(L m b_{m-1} y) .$$

Since $|\bar{n}[\omega_o^*] - n_o[\omega_o^*]| < 2 A b_{k_o}^{-1} < 2 A b_{k_o}^{-2}$ (6.40) yields

$$||S_{n_o}^*(x; \chi_F; \omega_o^*)| - |S_{\bar{n}}^*(x; \chi_F; \omega_o^*)|| \leq c_\mu\{C^*(\tilde{p}_o^*) + b_{k_o-1}y\} .$$

By (6.41) we have $C^*(\tilde{p}_o^*) < b_{m-1}y$. By combining results we obtain $|S_{n_o}^*(x; \chi_F; \omega_o^*)| = |S_{\bar{n}}^*(x; \chi_F; \bar{\omega}^*(x))| + O(L m b_{m-1}y)$.

Let $\omega_1^* = \bar{\omega}^*(x)$, $m_o = m$ and $n_1 = 4\cdot 2\pi\cdot \bar{n}[\omega_1^*]|\omega_1^*|^{-1}$.

The fact that $n_o[\omega_o^*] > 2 A b_{k_o}^{-2}$ is crucial in the proof of the following

(6.48) Lemma. $n_1 \leq (1 + b_{k_o})n_o$.

By (6.28), (6.42), (6.43) and (6.44) we have

$$|S_{n_o}^*(x; \chi_F; \omega_o^*)| = |S_{n_1}^*(x; \chi_F; \omega_1^*)| + O(L m_o b_{m_o-1}y) .$$

<u>Case 3.</u> $P_o^* \notin G^*_{k_oL}$ and $n_o[\omega_o^*] \leq 2 \; A \; b_{k_o}^{-2}$.

Choose $\tilde{n}$ as in (6.40) and $\overline{n},\overline{\omega}^*$, m as in (6.41). Then by (6.40)

$$||S^*_{n_o}(x;\chi_F;\omega_o^*)| - |S_o^*(x;\chi_F;\omega_o^*)|| \leq c_\mu\{C^*(\tilde{P}_o^*) + b_{k_o-1}y\}$$

But by (6.41) we have

$$||S^*_{n_o}(x;\chi_F;\omega_o^*)| - |S_o^*(x;\chi_F;\omega_o^*)|| \leq c_\mu\{b_{m-1}y + b_{k_o-1}y\} .$$

Clearly, if we let $m_o = m = 1$ we have

$$|S^*_{n_o}(x;\omega_o^*)| = |S_o^*(x;\omega_o^*)| + O(L\, m_o\, b_{m_o-1}y)$$

where it is understood that $\omega_o^* = \omega_1^*$.

We continue until Case (3) occurs or until Cases (1) and (2) yield an interval ω^*_{j+1} so small that $n_{j+1} = 0$.

APPENDIX A. THE HILBERT TRANSFORM

For f real valued with domain (a,b) and

$-\infty \leq a < x < b \leq \infty$. Let

$$(Hf)(x) = \text{P.V.}\int_a^b \frac{f(t)}{x-t}\,dt = \lim_{\varepsilon \to 0+} \left\{ \int_a^{x-\varepsilon} \frac{f(t)dt}{x-t} + \int_{x+\varepsilon}^b \frac{f(t)}{x-t}\,dt \right\}$$

(A.1) <u>Theorem.</u> If $f \in L^1(a,b)$, then $(Hf)(x)$ exists almost everywhere in (a,b).

(A.2) <u>Remark.</u> H is called the Hilbert transform of f over (a,b). If $f = u + iv$ where u and v are real valued with domain (a,b), then we define $(Hf)(x) = (Hu)(x) + i(Hv)(x)$.

For a proof of (A.1) see [15] page 132. In many of our applications f(t) will be the pointwise product of $g(t) = \varepsilon^{in\alpha t}$ and h(t) where $h \in L^1(a,b)$. Since $g \in L^\infty(a,b)$, we have of course by Holder's inequality that $f = gh \in L^1(a,b)$.

For an explanation of the notation used in the following theorem review the first section of chapter 5.

(A.3) <u>Theorem.</u> Let σ_x denote a subinterval of ω^* which contains x in its middle half. We define the maximal Hilbert transform H* of f over ω^*:

$$(H^*f)(x) = \sup_{\sigma_x} \left|\text{P.V.} \int_{\sigma_x} \frac{f(t)}{x-t}\,dt\right| \ ; \ x \in \omega^*; \ f \in L_\infty(\omega^*) .$$

Let $T = \{x \in \omega^* \mid (H^*f)(x) > y\}$; $y > 0$. Then

Part 1. $m(T) \leq c_\alpha |\omega^*| \varepsilon^{-c_\beta \frac{y}{\|f\|_\infty}}$

Part 2. For $\omega = \omega^*_{-1}$ $m(T) \leq C_p^p\, y^{-p} \|f\|_p^p$.

Proof of Part 1 (Taken directly from [3]) .

We may assume $\omega^* = (0,1)$ and $\|f\|_\infty = 1$. Extend $f(t)$ by 0 on $(1,2\pi)$ and denote by $E_o(t)$ the 2π-periodic extension of this function.

Lemma 1. If in Theorem (7.15) ([16] Vol 1 page 102) we suppose (in place of condition (7.17)) that

$$|F(x+t) + F(x-t) - 2F(x)| \leq A_1 t, \quad (0 < t < \pi, \ A_1 > 0, \ A_1$$

an absolute constant),

then

$$\left| \frac{\partial \tilde{F}(r,x)}{\partial x} - \left(-\frac{1}{\pi}\int^{\pi} (1-r) \frac{F(x+t)+F(x-t)-2F(x)}{4(\sin\frac{t}{2})^2} dt\right.\right| \leq A_2$$

$(0 < r < 1, \ A_2 > 0, \ A_2$ an absolute constant).

Proof. As indicated in Zygmund the estimate

$$Q'(r,t) - Q'(1,t) = 0(\delta_r^2 t^{-4}) \qquad (\delta_r = 1 - r)$$

is able to be stated precisely by

$$0 \leq Q'(r,t) - Q'(1,t) \leq A_1' \ \delta_r^2 \ t^{-4}$$

for $0 < r < 1$ and $0 < t < \pi$ where $A_1' > 0$ is an absolute constant, by considering the cases $0 < r \leq \frac{1}{2}$, $\frac{1}{2} < r < 1$.

For $0 < r \leq \frac{1}{2}$ we observe that $\Delta(r,t) \geq (1-r)^2 \geq \frac{1}{4}$

$(\Delta(r,t) = 1 - 2r\cos t + r^2)$, and for the other case we use the method of Zygmund.

Lemma. 2. If f is real valued, periodic of period 2π with $|f(x)| \leq 1$ for each x we have

$$\left|\tilde{f}(r,x) - \left(-\frac{1}{\pi}\int_{\delta r}^{\pi} \frac{f(x+t) - f(x-t)}{2 \operatorname{tg} \frac{t}{2}}\, dt\right)\right| \leq B_1$$

where $B_1 > 0$ is an absolute constant, $\delta_r = 1-r$, $0 < r < 1$.

Proof. By modifying f by a constant of absolute value ≤ 1, we are able to suppose that the indefinite integral F of f is periodic. Under these conditions

$$|F(x+t) + F(x-t) - 2F(x)| = \int_{x-t}^{x+t} |f(u)|du \leq B_1' t, \quad 0 < B_1'$$

an absolute constant, and $0 < t < \pi$.

We let $\Psi_x(t) = \dfrac{f(x+t)-f(x-t)}{2}$ and

$$\Psi(t) = F(x+t) + F(x-t) - 2F(x) .$$

Then

$$\int_{\delta r}^{\pi} \frac{\Psi(t)}{4(\sin \frac{t}{2})^2}\, dt = \int_{\delta r}^{\pi} \Psi(t)d\left(\frac{-1}{2 \operatorname{tg} \frac{t}{2}}\right) = \frac{\Psi(\delta_r)}{2 \operatorname{tg} \frac{\delta r}{2}} + \int_{\delta r}^{\pi} \frac{\Psi x(t)}{\operatorname{tg} \frac{t}{2}}\, dt$$

hence

$$\tilde{f}(r,x) - \left(-\frac{1}{\pi}\int_{\delta r}^{\pi} \frac{\Psi x(t)}{\operatorname{tg} \frac{t}{2}}\, dt\right) = \frac{\partial \tilde{F}(r,x)}{x} - \left(-\frac{1}{\pi}\int_{\delta r}^{\pi} \frac{\Psi x}{\operatorname{tg} \frac{t}{2}}\, dt\right) =$$

$$\left[\frac{\partial \tilde{F}(r,x)}{\partial x} - \left(-\frac{1}{\pi}\int_{\delta r}^{\pi} \frac{\Psi(t)}{4(\sin \frac{t}{2})^2}\, dt\right)\right] - \frac{1}{\pi} \frac{(\delta_r)}{2 \operatorname{tg} \frac{\delta r}{2}}$$

The absolute value of the square bracket is majorized, for $0 < r < 1$, by an absolute constant according to Lemma 1, and also the absolute value of the remaining term.

Lemma 3. If f is real valued, periodic of period 2π and such that $|f(x)| \leq 1$ for each x where $\tilde{f}(x)$ exists, then the expression

$$\sup_{\Delta} \left| \frac{1}{\pi} \int_{\Delta} \frac{f(x+t)}{2 \operatorname{tg} \frac{t}{2}} \, dt \right|$$

where the sup is taken relative to the intervals $\Delta \subset (-\pi,\pi)$ $0 \in \Delta$ for which the ratio of the distance of 0 to the extreme right and left of Δ lies between λ and $1/\lambda$ (where λ is fixed with $0 < \lambda < 1$) is majorized by

$$\sup_{\substack{\Delta \subset (-\pi,\pi) \\ \Delta \text{ centered about } 0}} \left| \frac{1}{\pi} \int_{\Delta} \frac{f(x+t)}{2 \operatorname{tg} \frac{t}{2}} \, dt \right| + c_\theta$$

where c_θ does not depend on λ .

Proof. Let $\Delta = (x_1, x_2)$ $(x_1 < 0,\ x_2 > 0$ with $0 < \lambda \leq \left|\frac{x_2}{x_1}\right| \leq \frac{1}{\lambda})$. Suppose for example $|x_1| < x_2$, we have

$$\left| - \frac{1}{\pi} \int_{\Delta} \frac{f(x+t)}{2 \operatorname{tg} \frac{t}{2}} \, dt \right| \leq \left| - \frac{1}{\pi} \int_{x_1}^{-x_1} \right| + \left| - \frac{1}{\pi} \int_{-x_1}^{x_2} \right|$$

with

$$\left| - \frac{1}{\pi} \int_{-x_1}^{x_2} \right| \leq B_1'' \int_{-x_1}^{x_2} \frac{1}{t} \, dt = B_1'' \log \left(\frac{x_2}{-x_1} \right) \leq B_1'' \log \left(\frac{1}{\lambda} \right)$$

where $B''_1 > 0$ is an absolute constant, where the result, for the case $|x_1| > x_2$ is treated in a similar manner.

Lemma 4. If f is real valued, periodic of period 2π with $|f(x)| \leq 1$ for each x where $\tilde{f}(x)$ exists, the expression

$$\sup_{\Delta} \left| -\frac{1}{\pi} \int_{\Delta} \frac{f(x+t)}{2 \operatorname{tg} \frac{t}{2}} \, dt \right|$$

where the sup is taken relative to the intervals $\Delta \subset (-\pi,\pi)$ $0 \in \Delta$ for which the ratio of the distances of 0 to the extreme right and left of Δ lies between λ and $1/\lambda$ (where λ is fixed with $0 < \lambda < 1$) is majorized by

$$c'_\alpha \sup_{0 < r < 1} |\tilde{f}(r,x)| + c'_\beta$$

where c'_α, c'_β do not depend on λ .

Proof. Free to choose the second constant an absolute constant majorizing the expressions

$$\left| -\frac{1}{\pi} \int_{x_1}^{-\frac{1}{3}} \frac{f(x+t)}{2 \operatorname{tg} \frac{t}{2}} \, dt \right| , \left| -\frac{1}{\pi} \int_{1/3}^{x_2} \frac{f(x+t)}{2 \operatorname{tg} \frac{t}{2}} \, dt \right| \quad (-\pi \leq x_2 < -\tfrac{1}{3}, \tfrac{1}{3} < x_2 \leq \pi)$$

we are able to suppose that $(x_1,x_2) = \Delta \subset (-\frac{1}{3}, \frac{1}{3})$. If Δ' is the interval contained in Δ having an extremity in common with Δ , the proof of Lemma 3 gives

$$\left| -\frac{1}{\pi} \int_{\Delta} \right| \leq \left| -\frac{1}{\pi} \int_{\Delta'} \right| + c''_\alpha$$

where c''_α does not depend on λ. If 2r is the length of Δ' $(0 < r < 1)$ we have

$$\left| -\frac{1}{\pi}\int_{\Delta'} \frac{f(x+t)}{2\ \mathrm{tg}\ \frac{t}{2}}\ dt \right| \leq |\tilde{f}(x)| + \left| -\frac{1}{\pi}\int_{1-r}^{\pi} \frac{f(x+t)-f(x-t)}{2\ \mathrm{tg}\ \frac{t}{2}}\ dt \right|$$

$$\leq \sup_{0<r'<1} |\tilde{f}(r',x)| + c''_\beta + \left| -\frac{1}{\pi}\int_{1-r}^{\pi} \frac{f(x+t)-f(x-t)}{2\ \mathrm{tg}\ \frac{t}{2}}\ dt \right|$$

according to Lemma 2 and the definition of $\tilde{f}(x)$. But

$$\left| -\frac{1}{\pi}\int_{1-r}^{\pi} \frac{f(x+t)-f(x-t)}{2\ \mathrm{tg}\ \frac{t}{2}}\ dt \right| \leq \left| \tilde{f}(r,x) - \left(-\frac{1}{\pi}\int_{1-r}^{\pi} \frac{f(x+t)-f(x-t)}{2\ \mathrm{tg}\ \frac{t}{2}} dt \right) \right| + |\tilde{f}(r,x)|$$

$$\leq B_1 + \sup_{0<r'<1} |\tilde{f}(r',x)|$$

by Lemma 2.

We now prove (A.3) in the following way:

$$\frac{1}{t} - \frac{1}{2\mathrm{tg}\ \frac{t}{2}} = O(1) \quad \text{for} \quad 0 < |t| < \pi$$

We infer that

$$H^*(x) \leq \sup_{\sigma_x} \left| \int_{\sigma_x} \frac{E_o(t)}{2\mathrm{tg}\frac{(x-t)}{2}}\ dt \right| + B_2\ ; \quad x \in (0,1)$$

where $0 < B_2$ is an absolute constant. Next, with Lemma 4 (and the condition on σ_x) we see that we have an absolute constant $B'_2 > 0$ such that

$$H^*(x) \leq B'_2 \sup_{0<r<1} |\tilde{E}_o(r,x)| + B'_2 .$$

We set, for the reason of simplification in writing, $k = \frac{\pi}{4}$.

We have (see[16] vol 1 page 254 Theorem (2.11))

$$\int_{-\pi}^{\pi} \varepsilon^{k|\tilde{E}_o(x)|}\,dx \le B_3 \qquad (0 < B_3 \text{ an absolute constant})$$

But then according to [16]vol 1 p. 155 Theorem 7.5 and that which is said after about the A_r and the Poisson kernel, we have, for $0 < r \le 1$

$$\int_{-\pi}^{\pi} \varepsilon^{k|\tilde{E}_o(r,x)|}\,dx = \int_{-\pi}^{\pi} \Big(\sum_{v=0}^{\infty} \frac{k^v}{v!}\,|\tilde{E}_o(r,x)|^v dx = \sum_{v=0}^{\infty} \int_{-\pi}^{\pi} \frac{k^v}{v!}\,|\tilde{E}_o(r,x)|^v dx$$

$$\le \sum_{v=0}^{\infty} A_v \int_{-\pi}^{\pi} \frac{k^v}{v!}\,|\tilde{E}_o(x)|^v dx \le A \int_{-\pi}^{\pi} \Big(\sum_{v=0}^{\infty} \frac{k^v}{v!}\,|\tilde{E}_o(x)\,|^v\Big)dx = A \int_{-\pi}^{\pi} \varepsilon^{k|\tilde{E}_o(x)\,|}\,dx$$

with the $A_v > 0$ which do not depend on v majorized by the absolute constant A. We have therefore

$$\int_{-\pi}^{\pi} \varepsilon^{k|\tilde{E}_o(r,x)|}\,dx \le A\,B_3 \quad (0 < r \le 1,\ A \text{ and } B_3 \text{ absolute constants } > 0)$$

Under these conditions, we consider the function h, analytic in the open unit disk, defined by

$$h(r,x) = \varepsilon^{\pm\frac{1}{2} k_i (E_o(r,x)+i\ \tilde{E}_o(r,x))} = \frac{1}{\pi} \int_{-\pi}^{\pi} \varepsilon^{\pm\frac{1}{2} k_i (E_o(t)+i\ \tilde{E}_o(t))}\, P(r,x-t)\,dt$$

P being the Poisson kernel. Since

$$\int_{-\pi}^{\pi} \Big|\, \varepsilon^{\pm\frac{1}{2} k_i (E_o(t)+i\ \tilde{E}_o(t))} \Big|^2 dt \le B_3$$

and since the Poisson kernel P satisfies the conditions of Lemma (7.1) in [16] vol 1 page 154, we have an absolute constant $B_3' > 0$ such that

$$\int_{-\pi}^{\pi} \sup_{0 < r < 1} \varepsilon^{\pm k \tilde{E}_o(r,x)} dx \leq B'_3$$

Hence

$$\int_{-\pi}^{\pi} \sup_{0 < r < 1} \varepsilon^{k|\tilde{E}_o(r,x)|} dx \leq 2 B'_3.$$

If we now set $k' = \frac{k}{B'_2}$, then

$$k' \; H^*(x) \leq k \sup_{0 < r < 1} |\tilde{E}_o(r,x)| + k \quad (x \in (0,1))$$

We deduce

$$\int_0^1 \varepsilon^{k'H^*(x)} dx \leq \int_{-\pi}^{\pi} \varepsilon^k \varepsilon^k \sup_{0 < r < 1} |\tilde{E}_o(r,x)| \, dx =$$

$$\int_{-\pi}^{\pi} \varepsilon^k \sup_{0 < r < 1} \varepsilon^{k|\tilde{E}_o(r,x)|} dx < 2 \, \varepsilon^k \, B'_3 = B''_3,$$

B''_3 being an absolute constant > 0. Moreover,

$$B''_3 \geq \int_{\{x \in (0,1) \, | H^*(x) \geq y\}} \varepsilon^{k'H^*(x)} dx$$

<u>Proof of Part 2.</u> (Taken directly from [8])

Let $\delta_x \subset \omega^*_{-1}$ denote an interval with center x, and define

$$(\bar{H}f)(x) = \sup_{\delta x} \left| \int_{\delta_x} \frac{f(t)}{x-t} dt \right|$$

and

$$(Hf)(x) = \sup_{\delta_x} \left| \int_{\omega^* - \delta_x} \frac{f(t)}{x-t} dt \right|$$

The Hardy-Littlewood maximal function of f is

$$\bar{f}(x) = \sup_{\delta_x} \frac{1}{|\delta x|} \int_{\delta x} |f(t)|\,dt$$

We have $\|\bar{f}\|_p < K_p \|f\|_p \quad ; \quad 1 < p < \infty$

But it is easy to see that

$$(H^*f)(x) \leq (\bar{H}f)(x) + c_\alpha \bar{f}(x) \leq 2(Hf)(x) + c_\alpha \bar{f}(x) \quad .$$

The result now follows from the corresponding result for the operator H.

APPENDIX B

Properties of $c_n(\omega)$, $C_n(\omega)$ and $S_n^*(\omega)$.

This section is taken directly from [8].

The word const. denotes an absolute constant.

An important property of the numbers $C_n(\omega; f)$ is that $C_n(\omega; f) = 0$ for some n if and only if $f(x) = 0$ for a.e. $x \in \omega$. This property is not shared with the numbers $c_n(\omega; f)$.

$C_n(\omega)$ and $c_n(\omega)$ are clearly related. Also, $C_n(\omega)$ is related to $S_n^*(\omega)$ as $c_n(\omega)$ is related to $S_n(\omega)$, the nth partial sum of the Fourier series of f over ω. That is, for $\omega = (0, 2\pi)$

$$S_n - S_{n-1} = c_n \text{ and } |S_n^*| - |S_{n-1}^*| = O(C_n).$$

Lemma (B.1) is the technical basis for the above relations.

(B.1.) Lemma Let $\phi(t) \in C^2(\omega)$, $|\omega| = 2\pi \cdot 2^{-\nu}$. Then we can represent $\phi(t)$.

$$(*) \quad \phi(t) = \Sigma \gamma_\mu \exp\{-i2^\nu \cdot 3^{-1} \cdot \mu t\}, \quad t \in \omega,$$

where $(1 + \mu^2)|\gamma_\mu| \le \text{const.} (\max_\omega |\omega| + 2^{-2\nu} \max_\omega |\phi''|)$.

Proof. By a change of variables, $t = 2^{-\nu}\tau$, we may assume $\omega = (0, 2\pi)$. We choose polynomials p_1, p_2 such that

$$\Psi(t) = \begin{cases} p_1(t) & [-2\pi, 0) \\ \phi(t) & (0, 2\pi) \\ p_2(t) & [2\pi, 4\pi] \end{cases} \quad \text{satisfies}$$

$\Psi \in C^2([-2\pi, 4\pi])$, $\Psi^{(k)}(-2\pi) = \Psi^{(k)}(4\pi) = 0$, $k = 0, 1, 2$. Then

$$\max_{[-2\pi,4\pi]} |\Psi| + \max_{[-2\pi,4\pi]} |\Psi''| \le \text{const}[\max_{(0,2\pi)} |\phi| + \max_{(0,2\pi)} |\phi''|]$$

The Fourier expansion of Ψ over $[-2\pi,4\pi]$ yields

$$\phi(t) = \Sigma \gamma_\mu \exp\{-i3^{-1} \mu t\}, \quad t \in \omega.$$

For $\mu = 0$ we have $|\gamma_\mu| \le \max |\Psi|$.

For $\mu \ne 0$ we integrate by parts two times to obtain

$$|\gamma_\mu| \le \frac{1}{\mu^2} \max |\Psi''|$$

(B.2) Lemma For any integer n and any $\omega = \omega_{j\nu}$ we have

$$|c_{n\cdot 2^{-\nu}}(\omega)| \le \text{const.} \; C_{n[\omega]}(\omega) \quad .$$

Proof. We apply Lemma (B.1) to the function $e^{-i\beta t}$, $\beta = n - n[\omega]\cdot 2^\nu$, $\omega = \omega_{j\nu}$. Since $0 \le \beta < 2^\nu$ we have $|\gamma_\mu| \le$ const. $(1+\mu^2)^{-1}$.

Then

$$c_{n\cdot 2^{-\nu}}(\omega) = \frac{1}{|\omega|}\int_\omega e^{-int} f(t) \; dt$$

$$= \frac{1}{|\omega|}\int_\omega e^{-i\beta t} \; e^{-in[\omega]2^\nu t} f(t) \; dt$$

$$= \sum_\mu \gamma_\mu \; c_{n[\omega]+\mu/3}(\omega)$$

Hence,

$$|c_{n\cdot 2^{\nu}}(\omega)| \le \text{const.} \; C_{n[\omega]}(\omega) \quad .$$

That is, if each $c_\alpha(\omega)$ is small, then each $C_n(\omega)$ is small. A more precise relation is given in Lemma (B.3).

(B.3) Lemma Given n and $g \in L^2(\omega)$. Suppose $\int_\omega |g(t)|^2 dt \leq G^2|\omega|$ and $|c_m(\omega)| \leq \mu$ whenever $|n-m| \leq M$ $(M \geq 2)$. Then $C_n(\omega; g) \leq \text{const.}\{\mu \log M + \frac{G}{M^{1/2}}\}$.

Proof. $C_n(\omega;g) = \frac{1}{10} \sum_{\mu=-\infty}^{\infty} |c_{n+\mu/3}(\omega)| \quad (1+\mu^2)^{-1}$.

$$c_{n+\mu/3}(\omega) = \frac{1}{|\omega|} \int_\omega g(x) \exp \{-2^\nu(n + \frac{\mu}{3})ix\} dx$$

$$= \frac{1}{|\omega|} \int_\omega g(x) \; e^{-i\alpha_\mu x} \; dx .$$

Let $e^{+i\alpha_\mu x} \sim \sum_k \alpha_{\mu,k} \, e^{i2^\nu xk}$. Then

$$c_{n+\mu/3}(\omega) = \sum_k c_k(\omega)\cdot\bar{\alpha}_{\mu,k} .$$

Also,

$$\alpha_{\mu,k} = \frac{1}{|\omega|} \int_\omega e^{i\alpha_\mu x} \cdot e^{-i2^\nu kx} dx$$

$$= \frac{1}{|\omega|} \int_\omega \exp\{-i2^\nu[k - (n + \frac{\mu}{3})]x\} dx .$$

If $k - (n + \frac{\mu}{3})$ is a non-zero integer, then $\alpha_{\mu,k} = 0$. If $k - (\alpha + \frac{\mu}{3}) = 0$ then $\alpha_{\mu,k} = 1$. Otherwise,

$$|\alpha_{\mu,k}| \leq \frac{2}{2\pi} |k - (n + \frac{\mu}{3})|^{-1} .$$

In any case,

$$|\alpha_{\mu,k}| \leq \text{const.} \; [1 + |k - n - \frac{\mu}{3}|]^{-1}$$

$$|c_{n+\mu/3}(\omega)| \leq \sum_k |c_k(\omega)| \, |\alpha_{\mu,k}|$$

$$\leq \text{const.} \; \sum_k |c_k(\omega)| \; [1 + |k-n- \frac{\mu}{3}|]^{-1} .$$

Suppose $|\frac{\mu}{3}| \leq \frac{1}{2}$ M. Then

$$\sum_{|k-n|\leq M} |c_k(\omega)|[1+|k-n-\frac{\mu}{3}|]^{-1} \leq \mu \sum_{|k-n|\leq M} [1+|k-n-\frac{\mu}{3}|]^{-1}$$

$$\leq \mu \sum_{|\ell|\leq M} \ell^{-1} \leq \mu(2+\log M)$$

Also,

$$\sum_{|k-n|>M} |c_k(\omega)|[1+|k-n-\frac{\mu}{3}|]^{-1}$$

$$\leq (\sum_k |c_k(\omega)|^2)^{1/2} (\sum_{|k-n|>M} [1+|k-n-\frac{\mu}{3}|]^{-2})^{1/2}$$

$$\leq G(\sum_{\ell>\frac{M}{2}} \ell^{-2})^{1/2} = \text{const.} \frac{G}{M^{1/2}} .$$

It follows that

$$|c_{n+\mu/3}(\omega)| \leq \text{const.} \ (\mu \log M + \frac{G}{M^{1/2}}) = q,$$

for $|\frac{\mu}{3}| \leq \frac{1}{2}$ M.

For $|\frac{\mu}{3}| \geq \frac{1}{2}$ M we use the estimate $|c_{n+\mu/3}(\omega)| \leq G$. Hence,

$$C_n(\omega;g) \leq \frac{1}{10} \sum_{|\frac{\mu}{3}|\leq\frac{M}{2}} |c_{n+\mu/3}|(1+\mu^2)^{-1} + \frac{1}{10} \sum_{|\frac{\mu}{3}|>\frac{M}{2}} |c_{n+\mu/3}|(1+\mu^2)^{-1}$$

$$\leq q(\frac{1}{10}) \sum_{\mu} (1+\mu^2)^{-1} + G \sum_{|\frac{\mu}{3}|>\frac{M}{2}} (1+\mu^2)^{-1}$$

$$\leq q + \text{const.} \ G\cdot M^{-1}$$

$$\leq \text{const.} \ q.$$

We will need the following property of the numbers $C_n(\omega)$.

(B.4) <u>Lemma</u> Suppose $g(x) = e^{i\lambda x}$ and $\omega=\omega_{\ell\nu}$. Then

$$|2^{-\nu}\lambda - n|\cdot C_n(\omega;g) \le \text{const.}$$

Proof. Since $|g(x)| = 1, |c_{n+\mu/3}(\omega;g)| \le 1$. It follows that $C_n(\omega;g) \le 1$. Hence, we may assume that $|2^{-\nu}\lambda - n| \ge 1$.

For $|\frac{\mu}{3}| \le \frac{1}{2}|2^{-\nu}\lambda - n|$ we have

$$|2^{-\nu}\lambda - n - \frac{\mu}{3}| \ge \frac{1}{2}|2^{-\nu}\lambda - n|$$

$$|c_{n+\mu/3}(\omega;g)| = |\frac{1}{|\omega|}\int_\omega e^{i\lambda x - i2^\nu (n+\mu/3)x}\, dx|$$
$$= |\frac{1}{|\omega|}\int_\omega \exp\{i2^{\nu}(2^{-\nu}\lambda - n + \frac{\mu}{3})x\}dx|.$$

If $2^{-\nu}\lambda - n + \frac{\mu}{3}$ ($\neq 0$) is an integer, then $|c_{n+\mu/3}(\omega;g)| = 0$.

In any case

$$|2^{-\nu}\lambda - n|\cdot|c_{\alpha+\mu/3}(\omega;g)| \le |2^{-\nu}\lambda - n|\cdot\frac{2}{2\pi}|2^{-\nu}\lambda - n + \frac{\mu}{3}|^{-1} \le \frac{2}{\pi}.$$

For $|\frac{\mu}{3}| > \frac{1}{2}|2^{-\nu}\lambda - n|$ we use $|c_{n+\mu/3}(\omega;g)| \le 1$. Then

$$|2^{-\nu}\lambda - n|C_n(\omega;g) \le |2^{-\nu}\lambda - n|\cdot\sum_{|\frac{\mu}{3}| \le \frac{1}{2}|2^{-\nu}\lambda - n|}|c_{n+\mu/3}(\omega;g)|(1+\mu^2)^{-1}$$
$$+ |2^{-\nu}\lambda - n|\cdot\sum_{|\frac{\mu}{3}| > \frac{1}{2}|2^{-\nu}\lambda - n|}|c_{n+\mu/3}(\omega;g)|(1+\mu^2)^{-1}$$
$$\le \frac{2}{\pi}\sum_\mu (1+\mu^2)^{-1} + |2^{-\nu}\lambda - n|\cdot\sum_{|\frac{\mu}{3}| \ge \frac{1}{2}|2^{-\nu}\lambda - n|}(1+\mu^2)^{-1}$$

$\le$ const.

$$S^*_n(x,\omega) = \int_{\omega^*}\frac{e^{-int} f(t)}{x-t}\, dt$$ corresponds to a partial sum of the Fourier series of f over ω^* of order $n[\omega^*]$. If

$|n_o[\omega^*] - n[\omega^*]| = 1$ then $S^*_n(x,\omega)$ and $S^*_{n_o}(x,\omega)$ differ by the modified Fourier coefficient $C^*_{n_o[\omega^*]}(\omega^*)$. This statement is made precise in Lemmas (B.5) and (B.6)

(B.5) <u>Lemma</u> Let $S^*_n(x,\omega) = \int_{\omega^*} \frac{e^{-int} f(t)}{x-t}\, dt$, where $\omega^* = \omega^*_{0\nu} = \omega_{0_\nu} \cup \omega_{1\nu}$. Suppose $2^{\nu+1} n_o[\omega^*] = n_o$. Then $|n-n_o| \leq 2^{\nu+1}$ implies

$$|e^{inx} S^*_n(x,\omega) - e^{in_o x} S^*_{n_o}(x,\omega)| \leq \text{const.} \max \{C_{n_o[\omega_{0_\nu}]}(\omega_{0\nu}), C_{n_o[\omega_{1\nu}]}(\omega_{0\nu})\}$$

<u>Proof.</u>

$$\Delta_o = e^{inx} \int_{\omega_{0\nu}} \frac{e^{-int} f(t)}{x-t}\, dt - e^{in_o x} \int_{\omega_{0\nu}} \frac{e^{-in_o t} f(t)}{x-t}\, dt$$

$$= \int_{\omega_{0\nu}} \{e^{in(x-t)} - e^{in_o(x-t)}\} \frac{f(t)}{x-t}\, dt$$

$$= \int_{\omega_{0\nu}} e^{in_o(x-t)} \{\phi(x-t)\} f(t)\, dt,$$

where $\phi(t) = \dfrac{e^{i(n-n_o)t} - 1}{t}$.

We can write $\phi(x-t) = \sum_\mu \gamma_\mu e^{i2^\nu \cdot 3^{-1} \mu(x-t)}$, $t \in \omega_{0\nu}$, where $(1+\mu^2)|\gamma_\mu| \leq \text{const.} \{\max_{\omega_{0\nu}} |\phi(x-t)| + 2^{-2\nu} \max_{\omega_{0\nu}} |\phi''(x-t)|\} \leq \text{const. } 2^\nu$

Hence

$$\Delta_o = \sum_\mu \gamma_\mu \int_{\omega_{0_\nu}} \exp\{i(n_o + 2^\nu \cdot 3^{-1} \cdot \mu)(x-t)\} f(t)\, dt$$

$$= \sum_\mu \gamma_\mu e^{i(n_o + 2^\nu \cdot 3^{-1}\mu)x} \int_{\omega_0} e^{-i2^\nu(2^{-\nu} n_o + \mu/3)t} f(t)\, dt$$

$$= \sum_\mu \gamma_\mu e^{i(n_o + 2^\nu \cdot 3^{-1}\mu)x} |\omega_{0\nu}|\, c_{2^{-\nu} n_o + \mu/3}(\omega_{o\nu}) .$$

Note that $2^{-\nu} n_o = 2n_o[\omega^*]$, an integer. Hence $2^{-\nu} n_o = n_o[\omega_{0\nu}]$.

Then

$$|\Delta_o| \le \Sigma |\gamma_\mu| \cdot |\omega_{0\nu}| |c_{n_o[\omega_{0\nu}]}(\omega_{0\nu})|$$

$$\le \text{const.} \quad C_{n_o[\omega_{0\nu}]}(\omega_{0\nu}) .$$

The integral over $\omega_{1\nu}$ is treated in the same way to complete the proof.

(B.6) Lemma Suppose $|\omega^*| = 4 \cdot 2\pi \cdot 2^{-(\nu+1)}$ and $2^{\nu+1} n_o[\omega^*] = n_o$.

Then $|n-n_o| \le 2^{\nu+1}$ implies

$$|e^{inx} S^*_n(x,\omega^*) - e^{in_o x} S^*_{n_o}(x,\omega^*)| \le \text{const.} \ C^*_{n_o[\omega^*]}(\omega^*) \quad .$$

Proof. Let $\omega^* = \omega_1 \cup \omega_2 \cup \omega_3 \cup \omega_4$ with $4|\omega_j| = |\omega^*|$, $j = 1,\ldots,4$. We integrate over each ω_j separately and use the fact that

$$C^*_{n_o}[\omega^*](\omega^*) = \max_{1 \le j \le 4} C_{n_o[\omega_j]}(\omega_j)$$

Let us consider the modified partial sums of the Fourier series of the function $f(x) = e^{inx}$ over ω^*.

(B.7) Lemma Let $S^*_n(x,\omega^*, e^{imx}) = \int_{\omega^*} \frac{e^{-int} e^{imt}}{x-t} dt$,

where m and n are integers and x is in the middle half of ω^* . Then

$$|S^*_n(x,\omega^*, e^{imx})| \le \text{const.}$$

Proof. Let $\bar{\omega}^*$ be the largest interval contained in ω^* with x the center of $\bar{\omega}^*$. Then

$$\left| \int_{\omega^*} \frac{e^{-int} \cdot e^{imt}}{x-t} \, dt \right| \leq \left| \int_{\bar{\omega}^*} \frac{e^{-int} \cdot e^{imt}}{x-t} \, dt \right| + 2 \log 3.$$

We may assume $x = 0$. Then

$$-i \int_{\bar{\omega}^*} \frac{e^{i(m-n)t}}{t} \, dt = \text{P.V.} \int_{\bar{\omega}^*} \frac{\sin(m-n)t}{t} \, dt$$

But

$$\lim_{\varepsilon \to 0+} \int_{\varepsilon}^{|\bar{\omega}^*|/2} \frac{\sin(m-n)t}{t} \, dt = \lim_{\varepsilon \to 0+} \int_{\varepsilon(m-n)}^{(m-n)|\bar{\omega}^*|/2} \frac{\sin t}{t} \, dt$$

$$= \int_0^{(m-n)|\bar{\omega}^*|/2} \frac{\sin t}{t} \, dt \leq \text{const.}$$

(B.8) Lemma. Let $\Omega = \{\omega_k\}$ be a partition of ω^* and let ω_k have midpoint t_k and length δ_k. We define

$$\Delta(x) = \Delta(x; \Omega) = \sum_k \frac{\delta_k^2}{(x-t_k)^2 + \delta_k^2} \qquad (x \in \omega^*)$$

Under these conditions if we set, for $M \geq 0$

$$U = U_M = \{x \in \omega^* \mid \Delta(x) > M\}$$

we have

$$m(U) = |U| \leq \text{const.}\, \varepsilon^{-\text{const.}\, M} |\omega^*| .$$

Proof. Taken directly from [3] . Without loss of generality we may assume $\omega^* = (0,1)$.

We set $\delta_k = (1-r_k)$ $(0 < r_k < 1)$. For $0 < x < 1$ we have (where P denotes the Poisson kernel)

$$\frac{\delta_k^2}{(x-t_k)^2+\delta_k^2} \leq \frac{\delta_k^2}{(x-t_k)^2 r_k^2+\delta_k^2} \leq \frac{\delta_k^2}{\left(2\sin\frac{(x-t_k)}{2}\right)^2 r_k^2+\delta_k^2}$$

$$\leq \delta_k^2 \frac{(1-r_k^2)}{\delta_k^2+r_k\left(2\sin\frac{(x-t_k)}{2}\right)^2} = 2\delta_k\, P(r_k, x-t_k) .$$

It is sufficient to establish the lemma by replacing $\Delta(x)$ by

$$\Delta_1(x) = \sum_k 2\delta_k \quad P(r_k, x - t_k)$$

We will utilize for this

The inequality of Harnack. This follows immediately:

With T designating the Torus of dimension one, let $g(x) \geq 0$ be a function of $L^1(T)$. Let $g(z) = g(r,x)$ be the corresponding harmonic function in the open unit disk $(z = r\,\varepsilon^{ix}$, $0 \leq r < 1)$, defined by

$$\begin{cases} g(r,x) = \frac{1}{\pi} \int_T g(x+t)\ P(r,t)\ dt \\ \frac{1}{2}\ \frac{(1-r)}{(1+r)} \leq P(r,t) \leq \frac{1}{2}\ \frac{(1+r)}{(1-r)} \qquad (0 \leq r < 1) \\ g(0) = \frac{1}{2\pi} \int_T g(t)dt \end{cases}$$

We have the following inequality of Harnack

$$\left(\frac{1-r}{1+r}\right) g(0) \le g(r,x) \le \left(\frac{1+r}{1-r}\right) g(0) \qquad (0 \le r < 1)$$

Corollary. Let $U(z)$ be harmonic and ≥ 0 in a domain D.

Let $z_o, z_o' \in D$ and suppose that for a $\lambda > 1$, the closed disk centered at z_o and of radius $\lambda|z_o - z_o'|$ be included in D. Then we have

$$U(z_o) \le \left(\frac{\lambda+1}{\lambda-1}\right) U(z_o')$$

which is essentially no different than the first inequality of Harnack.

We prove (B.8) by considering that (0,1) is on the torus T.

We form, for $g(x) \in L^1(T)$ with $g(x) \ge 0$ and $g(x) = 0$ for $x \notin (0,1)$ the following expression ($\Delta_1(x)$ having a sense on T).

$$\int_T \Delta_1(x) g(x) dx = \sum_k 2\,\delta_k \int_T g(x)\, P(r_k, x-t_k) dx =$$

$$\sum_k (2\pi\delta_k) \frac{1}{\pi} \int_T g(x)\; P(r_k, x-t_k) dx = 2\pi \sum_k \delta_k\, g(r_k, t_k)$$

if $g(r,x)$ $(0 \le r < 1)$ is the harmonic function corresponding to $g(x)$.

Since, for $t \in \omega_k$, the closed disk of center $r_k\, \varepsilon^{it_k}$ and of radius $2|r_k \varepsilon^{it} - r_k\, \varepsilon^{it_k}|$ is contained in the open unit disk, we have, according to the Corollary of the inequality of Harnack,

$$\delta_k\, g(r_k, t_k) = \int_{\omega_k} g(r_k, t_k) dt \le 3 \int_{\omega_k} g(r_k, t_k) dt \le 3 \int_{\omega_k} \sup_{0 \le r < 1} g(r,t)\, d$$

$$\le 3 \int_{\omega_k} \sup_{0 \le r < 1} g(r,t) dt$$

hence $\int_T \Delta_1(x) g(x)dx \leq \text{const.} \int_T \sup_{0 \leq r < 1} g(r,t)dt$

The last integral is majorized by an absolute constant if $\int_0^1 g(x) \log^+ g(x)dx$ remains itself majorized by an absolute constant. (See theorem on maximal functions in [16] vol. 1 page 155.)

We have also, $\Delta_1(x)$ being definable on T,

$$\int_0^1 \Delta_1(x)dx \leq \int_T \Delta_1(x)dx = \sum_k 2\delta_k \int_T P(r_k, x-t_k)dx$$

$$= (2\pi) \sum_k \delta_k \frac{1}{\pi} \int_T P(r_k, x-t_k)dx = 2\pi \sum_k \delta_k = 2\pi .$$

We consider, for $M \geq 0$, the set E_M defined by

$$E_M = \{x \in (0,1) \mid \Delta_1(x) > M\}$$

We have $|E_M| M \leq 2\pi$. We choose $M \geq 4\pi$ which implies $|E_M| \leq \frac{1}{2}$.

We put $|E_M| = \mu$ for $M \geq 4\pi$ (implies $0 \leq \mu \leq \frac{1}{2}$) and we suppose $\mu \neq 0$. (If $\mu = 0$, the inequality in vue is evident.)

We consider the function g_M defined by

$$g_M(x) = \begin{cases} \frac{1}{\mu \log\left(\frac{1}{\mu}\right)} & \text{if } x \in E_M \\ 0 & \text{otherwise} \end{cases}$$

Then we have

$$\int_0^1 g_M(x) \log^+ g_M(x)dx = \mu\left(\frac{1}{\mu \log \frac{1}{\mu}}\right) \log^+\left(\frac{1}{\mu \log \frac{1}{\mu}}\right) =$$

$$\frac{1}{(\log \frac{1}{\mu})} \left[\log \frac{1}{\mu} + \log\left(\frac{1}{\log \frac{1}{\mu}}\right)\right]$$

Since we have $\frac{1}{\mu \log \frac{1}{\mu}} > 1$ (which is the consequence of $\frac{1}{\mu} > \log \frac{1}{\mu}$).

Therefore $\int_0^1 g_M \log^+ g_M dx = 1 - \frac{\log\log \frac{1}{\mu}}{\log \frac{1}{\mu}} \quad O(1)$

for $0 < \mu \leq \frac{1}{2}$. Thus for $M \geq 4\pi$, we have

$$\int_T \Delta_1(x)\, g_M(x)dx \leq \text{const.}$$

hence

$$M \int_{E_M} g_M(x)dx = \frac{M}{\log \frac{1}{\mu}} \leq \int_T \Delta_1(x) g_M(x)dx \leq \text{const.}$$

and the result follows.

PROOF OF (6.11). (Taken directly from [8].)

If (n,ω) satisfies "A_k" then (n,ω) can be associated with a pair $(\lambda[\omega'],\omega') \in G_k$, where $\omega \subset \omega'$, $|\omega'| \leq b_k^{-10}|\omega|$. For each fixed pair $(\lambda[\omega'],\omega') \in G_k$ there are < const. $\log b_k^{-1}$ different dyadic lengths $|\omega|$ which satisfy $|\omega'| \geq |\omega| \geq b_k^{10}|\omega'|$.
For each different length there are < const. b_k^{-10} integers n which satisfy $|n - \lambda[\omega]| < b_k^{-10}$. It follows that

$$\Sigma|\omega| \leq \text{const. } b_k^{-11} |\omega'| ,$$

where the sum is taken over all pairs (n,ω) which are associated with the fixed pair $(\lambda[\omega'], \omega') \in G_k$ and which satisfy "A_k".

Then

$$\sum_{A_k}|\omega| < \text{const. } b_k^{-11} \sum_{G_k} |\omega'| \leq \text{const. } b_k^{-13} y^{-p} m(F).$$

If (n,ω) satisfies "B_k" then $\omega \subset \omega'$ for some ω' such that $(n',\omega') \varepsilon\ G_k$. For each fixed $(n',\omega')\ \varepsilon\ G_k$, $P_k(x;\omega')$ contains $\leq b_k^{-3}$ exponents λ, so the number of pairs (λ,λ') corresponding to $P_k(x;\omega')$ is $\leq b_k^{-6}$. For each fixed pair (λ,λ'), $\lambda \neq \lambda'$, $b_k^{10} \leq |\lambda-\lambda'|\cdot|\omega| \leq b_k^{-20}$ holds for $<$ Const. $\log b_k^{-1}$ choices of dyadic lengths $|\omega|$. For each fixed ω, $|n - \lambda[\omega]| < b_k^{-10}$ holds for $<$ const. b_k^{-10} different integers n. It follows that for fixed $(n',\omega')\ \varepsilon\ G_k$,

$$\Sigma\ |\omega| \leq \text{const. } b_k^{-17}\ |\omega'|$$

where the sum is taken over all pairs (n,ω) which satisfy "B_k" and $\omega \subset \omega'$.

Hence

$$\sum_{"B_k"} |\omega| \leq \text{const. } b_k^{-17} \sum_{G_k} |\omega'| \leq \text{const. } b_k^{-19} y^{-p}\ m\ (F)$$

combining the contribution from "A_k" and "B_k" the lemma follows.

<u>PROOF OF (6.20).</u> (Taken directly from [8].)

If $\omega^* \not\subset S^*$, then $\omega' \not\subset S$ for each ω' such that $4|\omega'| = |\omega|$ and $\omega' \subset \omega^*$. Hence by the definition of S we have

$$C_n(\omega') \leq \frac{1}{|\omega'|} \int_{\omega'} \chi^o_F(x)dx < y^p$$

for each such ω'. Consequently, $C^*_n(\omega^*) < y^p$.

So that $b_k y < y^p$ or $y > b_k^{1/(p-1)}$

Suppose $1 < p \leq 2$. Then $(p/2-1) \leq 0$; so that

$$y^{(p/2-1)} \leq b_k^{(p-2)/2(p-1)} = b_{kL_1}^{-(2-p)/2L_1(p-1)}$$

Suppose we can choose the (large) positive integer $L_1 = L_1(p)$ such that $\frac{(2-p)}{2L_1\{p-1\}} \leq \frac{1}{4}$. Then we have $y^{p/2} \leq b_{kL_1}^{-1/4} y$.

Suppose $2 \leq p < \infty$. We know

$$C^*_n(\omega^*) \leq 1$$

so that $b_k y \leq 1$

hence $y \leq b_k^{-1}$. Since $((p/2) - 1) \geq 0$ we have

$$y^{(p/2)-1} \leq b_k^{-(p-2)/2} = b_{kL_2}^{-(p-2)/2L_2}$$

Suppose we choose the (large) positive integer $L_2 = L_2(p)$ such that

$$\frac{(p-2)}{2\ L_2} \leq 1/4$$

Then $y^{p/2} \leq b_{k\ L_2}^{-1/4}\ y$. Let $L = L(p) = L_1(p) + L_2(p)$.

PROOF OF 6.31. (Taken directly from [9].)

By (B.2) we have $|E_n(t)| = c_{n\cdot 2}{}^{-\mu(\omega_{j\mu})}| \leq$ Const. $C_{n[\omega_{j\mu}]}(\omega_{j\mu})$.

But (6.21) implies $|E_n(t)| \leq$ Const. $b_{k-1}y$; $t \in \omega^*$.

PROOF OF 6.32.

Immediate by (A.3) Part 1 and (B.8).

PROOF OF (6.33). (Taken directly from [9]).

We write

$$\int_{\omega_o^*-\omega^*(x)} \frac{\varepsilon^{-int} \chi_F^o(t)}{x-t}\, dt = H_n(x) + R_n(x)$$

where

$$H_n(x) = \int_{\omega_o^* - \omega^*(x)} \frac{E_n(t)}{x-t}\, dt$$

$$R_n(x) = \int_{\omega_o^*-\omega^*(x)} \frac{\varepsilon^{-int} \chi^o_F(t) - E_n(t)}{x-t}\, dt$$

$H_n(x)$ is majorized by $2\, H_n^*(x)$, where $H_n^*(x)$ is the maximal Hilbert transform of $E_n(t)$ over ω^*.

Denote by δ_j the length of ω_j and by t_j the midpoint of ω_j for each $\omega_j \varepsilon \Omega(p^*;k)$. $\omega_o - \omega^*(x)$ is the union of a certain subset (x) of the intervals $\omega_j \varepsilon \Omega(p^*;k)$. Using the fact that the numerator in the integrand of $R_n(x)$ has vanishing integral over each ω_j, we write

$$R_n(x) = \sum_{(x)} \int_{\omega_j} \frac{(t-t_j)}{(x-t)\,(x-t_j)}\, \varepsilon^{-int} \chi^o_F(t)\, dt$$

$$- \sum_{(x)} \int_{\omega_j} \frac{(t-t_j)}{(x-t)\,(x-t_j)}\, E_n(t)\, dt$$

(6.30) and (6.31) imply that the second term on the right above is dominated by const. $b_{k-1} y \cdot \sum_{(x)} \left(\delta_j^2 \Big/ (x-t_j)^2 + \delta_j^2 \right)$. To obtain this same estimate for the first term on the right above we introduce the function

$$\phi(t) = \frac{(t-t_j)}{(x-t)(x-t_j)} \varepsilon^{-i(n-2^{\nu}n[\omega_j])t}, \quad t \in \omega_j, \ |\omega_j| = 2\pi \cdot 2^{-\nu} .$$

Note that $x \notin \omega_j$ so $\phi \in C^2(\omega_j)$. According to (B.1) we write

$$\phi(t) = \Sigma \gamma_{\mu} \exp\{-i2^{\nu} \cdot 3^{-1} \mu t\}, \quad t \in \omega_j,$$

where $(1 + \mu^2) |\gamma_{\mu}| \leq \text{Const.} \ (\delta_j / ((x-t_j)^2 + \delta_j^2))$. (The last inequality follows from the fact that $|t-x| \geq \delta_{j/2}$ for $t \in \omega_j$ and $|n-2^{\nu}n[\omega_j]| \leq 2^{\nu}$.) With appropriate substitution this yields

$$\left| \sum_{(x)} \int_{\omega_j} \frac{t-t_j}{(x-t)(x-t_j)} \varepsilon^{-int} \chi_F^o(t)dt \right| \leq \text{Const.} \ \sum_{(x)} \frac{\delta_j^2}{(x-t_j)^2+\delta_j^2} C_{n[\omega_j]}(\omega_j)$$

$$\leq \text{Const.} \ b_{k-1} y \sum_{(x)} \frac{\delta_j^2}{(x-t_j)^2+\delta_j^2}$$

By adding (positive) terms of $\Delta(x)$ corresponding to the remaining intervals of $\Omega(p^*;k)$ we obtain

$$|R_n(x)| \leq \text{Const.} \ b_{k-1} \ \Delta(x).$$

PROOF OF 6.36

Immediate by (A.3) part 2.

PROOF OF 6.40. (Taken directly from [9].)

We first investigate the polynomial $P_k(x;\omega)$ under the condition $(n,\omega) \notin \tilde{G}_k$ $n > 0$ (and $\omega \notin X_k$).

We write

$$P_k(x;\omega) = Q_o(x;\omega) + Q_1(x;\omega)$$

where $Q_1(x,\omega)$ contains the terms $a \, \varepsilon^{i\lambda x}$ of $P_k(x;\omega)$ for which

$|n - \lambda[\omega]| \geq b_k^{-10}$.

If λ and λ' are exponents of terms of $Q_o(x;\omega)$, then "B_k" implies $|\lambda-\lambda'|\cdot|\omega| < b_k^{10}$. This and (6.6) can be used to show

$$Q_o(x;\omega) = \rho\varepsilon^{i\lambda x} + O(b_k^8\, y^{p/2}), \quad x \text{ near } \omega$$

where ρ is constant and λ is an exponent of $Q_o(x;\omega)$.

If $(n, \omega^*) \notin G_k^*$ and $\omega^* \not\subset X_k^*$ we can write

$$P_k(x;\, \omega') = \rho'\, \varepsilon^{i\lambda' x} + O(b_k^8\, y^{p/2}) + Q_1(x;\omega')$$

for each of the four intervals $\omega' \subset \omega^*$, $4|\omega'| = |\omega^*|$. For our purposes, the essential part of $P_k(x;\omega')$ is the term $\rho'\, \varepsilon^{i\lambda' x}$. If $(n,\omega^*) \notin G_k^*$ and $\omega^* \not\subset X_k^* \cup Y_k^*$, then the polynomials $Q_o(x;\omega')$ corresponding to the four subintervals of ω^*, $4|\omega'| = |\omega^*|$, are identical. To see this suppose $P_k(x;\omega_o')$ contains a term $a\,\varepsilon^{i\lambda x}$ with $|\lambda[\omega_o'] - n| < b_k^{-10}$. Then $(\lambda[\bar{\omega}],\bar{\omega}) \in G_k$ for some $\bar{\omega} \supset \omega_o'$. Since $(n,\omega_o') \notin \tilde{G}_k$, "$A_k$" implies $|\bar{\omega}| > b_k^{-10}|\omega_o'|$. If $\bar{\omega}$ does not contain each ω', then ω' would be contained in Y_k^*. Hence, $\bar{\omega} \supset \omega'$ for each of the four ω', so a $\varepsilon^{i\lambda x}$ is a term of each $P_k(x;\omega')$. Hence the terms $\rho\,\varepsilon^{i\lambda x}$ may be chosen the same for each of the four intervals of ω^*.

Now let ω_o' be the subinterval of ω_o^* for which

$C_{n_o}[\omega_o'](\omega_o') = C^*(P_o^*)$ and ω' be any interval $\omega' \subset \omega_o^*$, $4|\omega'| = |\omega_o'|$.

Let P_o and P be the corrsponding (kL)-polynomials.

We have $\left(1/|\omega'| \int_{\omega'} |\chi_F^o - P|^2 dx\right)^{1/2} \leq (1 + b_{kL}^{-2}) y^{p/2}$ and $|c_m(\omega';\chi_F - P)| \leq b_{kL}\, y^{p/2}$ for all m. (B.3) (with $M = b_{kL}^{-10}$)

and the estimate $y^{p/2} \leq b_{kL}^{-1/4}\, y$ then yield

(B.9) $C_n(\omega';\chi_F - P) \leq b_{kL}^{1/2}\, y$ for all n.

In particular, with $n = n_o[\omega_o']$, $\omega' = \omega_o'$, the above equation yields

(B.10) $C_{n_o[\omega_o']}(\omega_o'; P_o) \geq (b_k - b_{kL}^{1/2})\, y$

If every exponent λ of P_o satisfied $|\lambda[\omega_o'] - n_o[\omega_o']| \geq b_{kL}^{-5}$, we would have

$$C_{n_o}[\omega_o'](\omega_o';P_o) \leq \text{Const. } b_{kL}^{5}\, \Sigma|a_n| \leq \text{Const. } b_{kL}^{3}\, y^{p/2} \leq b_{kL}^{2}\, y,$$

a contradiction to (B.10). It follows that P_o (and hence each P) contains an exponent λ with $|\lambda[\omega_o'] - n_o[\omega_o']| < b_{kL}^{-5}$. Set $\tilde{n} = \lambda$ for such a λ. Then $(\tilde{n}[\tilde{\omega}'],\tilde{\omega}') \in G_{kL}$ for some $\tilde{\omega}' \supset \omega_o^*$. Take $\tilde{\omega}'$ with three of its neighbors to form $\tilde{\omega}^*$ with x in it's middle half. $\tilde{\omega}^* \supset \omega_o^*$, $4\cdot 2\pi \cdot \tilde{n}[\tilde{\omega}^*] = \tilde{n}|\tilde{\omega}^*|$ and $\tilde{P}^* \in \tilde{G}_{kL}^*$.

Since $p_o^* \notin G_{kL}^*$, we can write $P = \rho\varepsilon^{i\tilde{n}x} + Q_o'(x) + Q_1(x)$,

where $Q_1(x)$ contains only exponents λ' of P with $|\lambda'[\omega_o'] - n_o[\omega_o']| \geq b_{kL}^{-10}$ and $Q_o'(x) = O(b_{kL}^{7}\, y)$. Hence

$$C_{n[\omega']}(\omega'; P - \rho\varepsilon^{i\tilde{n}x}) = O(b_{kL}^{7}\, y) \text{ for } |n[\omega'] - n_o[\omega']| < b_{kL}^{-9}.$$

(B.9) and (B.10) can then be replaced by

(B.11) $C_{n[\omega']}(\omega';\chi_F - \rho\varepsilon^{i\tilde{n}x}) \leq \text{Const } b_{kL}^{1/2}\, y, |n[\omega'] - n_o[\omega']| \leq b_{kL}^{-9}$,

and

(B.12) $C_{n_o[\omega_o']}(\omega_o'; \rho\varepsilon^{i\tilde{n}x}) \geq \text{Const. } b_k y$.

(B.11) with $n = \tilde{n}$ yields

(B.13) $|\rho| \leq \text{const } C_{\tilde{n}[\omega']}(\omega'\rho\varepsilon^{i\tilde{n}x}) \leq \text{const } (C^*(\tilde{P}^*) + b_{kL}^{1/2}\, y)$.

In particular, $|\rho| \leq$ const. y . From (B.12) we have

$$\text{const } b_k y \leq |\rho| \text{ const } |n_o[\omega'_o] - \tilde{n}[\omega'_o]|^{-1} ,$$

or

$$|\tilde{n}[\omega'_o] - n_o[\omega'_o]| \leq \text{Const } b_k^{-1} y^{-1} |\rho| \leq A b_k^{-1} .$$

Suppose n satisfies $|n[\omega^*_o] - n_o[\omega^*_o]| < 2 \ A b_k^{-2}$. Write

$$|\varepsilon^{inx} S^*_n(x;\omega^*_o;\chi_F) - \varepsilon^{in_o x} S^*_{n_o}(x;\omega^*_o;\chi_F)|$$

$$\leq |\varepsilon^{inx} S^*_n(x;\omega^*_o;\chi_F - \rho\varepsilon^{i\tilde{n}x}) - \varepsilon^{in_o x} S^*_{n_o}(x;\omega^*_o;\chi_F - \rho\varepsilon^{i\tilde{n}x})|$$

$$+ |S^*_n(x;\omega^*_o;\rho\,\varepsilon^{i\tilde{n}x})| + |S^*_{n_o}(x;\omega^*_o;\rho\,\varepsilon^{i\tilde{n}x})| .$$

According to (B.7) and (B.13), each of the last two terms are majorized by const $|\rho| \leq$ const $\{C^*(P^*_o) + b_{kL}^{1/2} y\}$. The first term is estimated by using (B.11) and applying (B.6) $\leq 2 A b_k^{-2}$ times. The resulting bound is $\leq$ Const $2 \ A b_k^{-2} b_{kL}^{1/2} y < b_{k-1} y$. Combining results we obtain the desired estimate.

<u>PROOF OF 6.41.</u> (Taken directly from [9].)

Let Σ denote the collection of all triples (n,ω^*,ℓ), where

(i) $\omega^* \supset \omega^*_o$, x belongs to the middle half of ω^* and

$$4\cdot 2\pi\cdot n[\omega^*] = n|\omega^*|;$$

(ii) $1 \leq \ell \leq k$ and $C^*(\tilde{P}^*_o) < b_{\ell-1} y$;

(iii) $|n[\omega^*_o] - n_o[\omega^*_o]| \leq A \sum_{j=1}^{k} b_j^{-1}$; and

(iv) $(n[\omega^*],\omega^*) \in G^*_{\ell L}$

We must show that Σ is nonempty.

(A) If $C^*(\tilde{P}^*_o) < b_{k-1} y$, then $(\tilde{n}, \tilde{\omega}^*, k) \in \Sigma$. ($\tilde{n}$ and $\tilde{\omega}^*$ are as in (6.40)).

(A') If $C^*(\tilde{P}^*_o) \geq b_{k-1} y$, we define ℓ, $1 \leq \ell < k$, by $b_\ell y \leq C^*(\tilde{P}^*_o) < b_{\ell-1} y$.

(a) If $\tilde{P}^*_o \in G^*_{\ell L}$, then $(\tilde{n}', \omega^*_o, \ell) \in \Sigma$, $\tilde{n}' = 4\cdot 2\pi\, \tilde{n}[\omega^*_o]|\omega^*_o|^{-1}$.

(a') If $\tilde{P}^*_o \notin G^*_{\ell L}$ we apply (6.40) with n_o and k replaced by $\tilde{n}'$ and ℓ. We obtain a new pair $\tilde{n}_1, \tilde{\omega}^*_1$ such that $(\tilde{n}_1, \tilde{\omega}^*_1, \ell) \in \Sigma$. Note that

$$
\begin{aligned}
|\tilde{n}_1[\omega^*_o] - n_o[\omega^*_o]| &\le |\tilde{n}_1[\omega^*_o] - \tilde{n}'[\omega^*_o]| + |\tilde{n}'[\omega^*_o] - n_o[\omega^*_o]| \\
&\le |\tilde{n}_1[\tilde{\omega}^*_1] - \tilde{n}'[\tilde{\omega}^*_1]| + |\tilde{n}'[\omega^*_o] - n_o[\omega^*_o]| \\
&\le A \sum_{j=\ell}^{k} b_j^{-1}
\end{aligned}
$$

This proves Σ is nonempty.

Choose $(\bar{n}, \bar{\omega}^*, m) \in \Sigma$ such that m is minimal. The conclusions of our lemma are clear, except for the inequality $C^*(\bar{p}^*) < b_{m-1} y$ and the statements concerning $\Omega(\bar{p}^*; m)$.

Let us assume $C^*(\bar{p}^*) \ge b_{m-1}y$. We then define ℓ, $1 \le \ell < m$, by $b_\ell y \le C^*(\bar{p}^*) < b_{\ell-1}y$. If $\bar{p}^* \in G^*_{\ell L}$, then $(\bar{n}, \bar{\omega}^*, \ell) \in \Sigma$, a contradiction to the fact that m is minimal. If $\bar{p}^* \notin G^*_{\ell L}$, we apply (6.40) with ω^*_o, n_o, k replaced by $\bar{\omega}^*$, $\bar{n}$, ℓ. We obtain a new pair $\tilde{n}_1$, $\tilde{\omega}^*_1$ such that $(\tilde{n}_1, \tilde{\omega}^*_1, \ell) \in \Sigma$, a contradiction since $\ell < m$. It follows that $C^*(\bar{p}^*) < b_{m-1} y$.

We can now assert that the partition $\Omega(p^*; m)$ is defined. Let $\bar{\omega}^*(x)$ correspond to this partition and then set $\bar{p}^*(x) = (\bar{n}[\bar{\omega}^*(x)], \bar{\omega}^*(x))$. Note that x is in the middle half of ω^*_o and $\bar{\omega}^*(x)$. Since $x \notin W^*$, this implies $\bar{\omega}^*(x) \supset \omega^*_o$ or $\bar{\omega}^*(x) \subset \omega^*_o$ (strictly). Suppose $\bar{\omega}^*(x) \supset \omega^*_o$. Then $|\bar{\omega}^*(x)| > 2\cdot 2\pi 2^{-N}$, so $C^*(\bar{p}^*(x)) \ge b_{m-1}y$, (see 6.29). This implies $b_\ell\, y \le C(\bar{p}^*(x)) < b_{\ell-1}y$ for some $1 \le \ell < m$. If $\bar{p}^*(x) \in G^*_{\ell L}$, then $(\bar{n}, \bar{\omega}^*(x), \ell) \in \Sigma$, a contradiction. If $\bar{p}^*(x) \notin G^*_{\ell L}$, we can use (6.40) as before to obtain a contradiction. Hence $\bar{\omega}^*(x) \subset \omega_o^*$ (strictly).

Since $\bar{\omega}^*(x)$ is a union of intervals of $\Omega(\bar{p}^*;m)$ it remains only to prove that ω_o^* is a union of intervals of $\Omega(\bar{p}^*;m)$. This follows from the fact that $\bar{\omega}^*(x) \subset \omega_o^*$ (strictly) and by the construction of $\bar{\omega}^*(x)$.

PROOF OF 6.42.

This follows immediately from the fact that

$$\frac{1 + (v-3)^2}{(1+v)^2} \quad \text{const which is independent of } v.$$

PROOF OF 6.43.

This follows immediately from (B.5).

PROOF OF (6.47). (Taken directly from [9].)

Suppose not. Then we have $|c_m(\omega_{10})| < b_{kL} y^{P/2}$ for all m such that $|m-n| < b_{kL}^{-10}$. Since

$$\left(\left(1/_{|\omega_{10}|}\right) \int_{\omega_{10}} |\chi_F^o(x)|^2 dx\right)^{1/2} < y^{P/2}, \quad \text{we can use (B.3)}$$

(with $M = b_{kL}^{-10}$) and the estimate $y^{p/2} \leq b_{kL}^{-1/4}\, y$ to obtain

$C_n(\omega_{10}) < b_{kL}^{1/2}\, y < b_k y$, a contradiction.

PROOF OF (6.48).

If $\bar{n} \leq n_o$ then trivial. Suppose $n_o < \bar{n}$. Then since $2\ A\ b_{k_o}^{-1}$ is an integer we have $\left(\frac{\bar{n}}{2^{v+1}} - \frac{n_o}{2^{v+1}}\right) < 2\ A\ b_{k_o}^{-1}$. But since $2\ A\ b_{k_o}^{-2} < n_o[\omega_o^*]$ we have $2\ A\ b_{k_o}^{-2} < \frac{n_o}{2^{v+1}}$; so that

$$2\ A\ b_{k_o}^{-1} < \frac{n_o}{2^{v+1}}\, b_{k_o}.$$

Hence

$$\left(\frac{\bar{n}}{2^{v+1}} - \frac{n_o}{2^{v+1}} \right) < \frac{n_o}{2^{v+1}} \quad b_{k_o}$$ which implies

$$\bar{n} < \left(1 + b_{k_o} \right) n_o \ .$$

APPENDIX C. THE RESULTS OF KAHANE AND KATZNELSON

(C.0) We denote by R the additive group of real numbers and by Z the subgroup consisting of the integers. The group T is defined as the quotient $R/2\pi Z$ where, as indicated by the notation, $2\pi Z$ is the group of the integral multiples of 2π. There is an obvious identification between functions on T and 2π- periodic functions on R, which allows an implicit introduction of notions such as continuity, differentiability, etc. for functions on T. The Lebesgue measure on T also can be defined by means of the preceding identification: a function f is integrable on T if the corresponding 2π-periodic function, which we denote again by f, is integrable on $[0,2\pi]$ and we set

$$\int_T f(t)dt = \int_0^{2\pi} f(x)dx.$$

In other words, we consider the interval $[0,2\pi]$ as a model for T and the Lebesgue measure dt on T is the restriction of the Lebesgue measure of R to $[0,2\pi]$. The total mass of dt on T is equal to 2π and many of our formulas would be simpler if we normalized dt to have total mass 1, that is, if we replace it by $dx/2\pi$. Taking intervals on R as "models" for T is very convenient, however, and we choose to put $dt = dx$ in order to avoid confusion. We "pay" by having to write the factor $1/2\pi$ in front of every integral.

An all-important property of dt on T is its translation invariance, that is, for all $t_0 \in T$ and f defined on T,

$$\int f(t-t_0)dt = \int f(t)dt$$

We denote by $L^1(T)$ the space of all complex-valued, Lebesgue integrable functions on T. For $f \in L^1(T)$ we put

$$\|f\|_{L^1} = \frac{1}{2\pi} \int_T |f(t)|dt,$$

It is well known that $L^1(T)$, with the norm so defined, is a Banach space.

(C.1) We consider a homogeneous Banach space B on T.

DEFINITION: A set $E \subset T$ is a set of divergence for B if there exists an $f \in B$ whose Fourier series diverges at every point of E.

(C.2) DEFINITION: For $f \in L^1(T)$ we put

$$S_n^*(f,t) = \sup_{m \leq n} |S_m(f,t)|$$

(c.1)

$$S^*(f,t) = \sup_n |S_n(f,t)|$$

THEOREM: E is a set of divergence for B if, and only if, there exists an element $f \in B$ such that

(c.2) $S^*(f,t) = \infty$ for $t \in E$.

The theorem is an easy consequence of the following:

LEMMA: Let $g \in B$; then there exist an element $f \in B$ and a positive even sequence $\{\Omega_j\}$ such that $\Omega_j \to \infty$ monotonically with j and $\hat{f}(j) = \Omega_j \hat{g}(j)$ for all $j \in Z$, where $\hat{f}(j)$ denotes the jth complex Fourier coefficient for f.

PROOF OF THE LEMMA: For each n let $\lambda(n)$ be such that $\|\sigma_{\lambda(n)}(g) - g\|_B < 2^{-n}$. We write $f = g + \sum_{n=1}^{\infty} (g - \sigma_{\lambda(n)}(g))$.

The series defining f converges in norm; hence $f \varepsilon B$. Also $\hat{f}(j) = \Omega_j \hat{g}(j)$ where $\Omega_j = 1 + \Sigma_{n=1}^{\infty} \min(1, |j|/(\lambda(n) + 1))$.

PROOF OF THE THEOREM: The condition (c.2) is clearly sufficient for the divergence of $\Sigma\, \hat{f}(j)e^{ijt}$ for all $t \varepsilon E$. Assume, on the other hand, that for some $g \varepsilon B$, $\Sigma \hat{g}(j)e^{ijt}$ diverges at every point of E. Let $f \varepsilon B$ and $\{\Omega_j\}$ be the function and the sequence corresponding to g by the lemma. We claim that (c.2) holds for f. This follows from: $n > m$,

$$S_n(g,t) - S_m(g,t) = \sum_{m+1}^{n} (S_j(f,t) - S_{j-1}(f,t))\Omega_j^{-1}$$

(c.3)
$$= S_n(f,t)\Omega_n^{-1} - S_m(f,t)\Omega_{m+1}^{-1} + \sum_{m+1}^{n-1} (\Omega_j^{-1} - \Omega_{j+1}^{-1})\, S_j(f,t),$$

hence

$$|S_n(g,t) - S_m(g,t)| \leq 2S^*(f,t)\, \Omega_{m+1}^{-1}.$$

It follows that if $S^*(f,t) < \infty$, the Fourier series of g converges and $t \notin E$.

Remark: Let $\{\omega_n\}$ be a sequence of positive numbers such that $\omega_j = O(\Omega_j)$, $\Sigma_1^{\infty} (\Omega_j^{-1} - \Omega_{j+1}^{-1})\, \omega_j < \infty$; then, for all $t \varepsilon E$, $S_j(f,t) \neq o(\omega_j)$. This follows immediately from (c.3).

(C.3) For the sake of simplicity we assume throughout the rest of this section that (c.4) If $f \in B$ and $n \in Z$ then $e^{int} f \in B$ and $\|e^{int} f\|_B = \|f\|_B$

LEMMA: Assume (c.4); then E is a set of divergence for B if, and only if, there exists a sequence of trigonometric polynomials $P_j \in B$ such that $\Sigma \|P_j\|_B < \infty$ and

(c.5) $$\sup_j S^*(P_j, t) = \infty \quad \text{on } E.$$

PROOF: Assume the existence of a sequence $\{P_j\}$ satisfying $\Sigma \|P_j\|_B < \infty$ and (c.5). Denote by m_j the degree of P_j and let v_j be integers satisfying

$$v_j > v_{j-1} + m_{j-1} + m_j.$$

Put $f(t) = \Sigma e^{iv_j t} P_j(t)$. For $n \le m_j$ we have

$$S_{v_j+n}(f,t) - S_{v_{j-n-1}}(f,t) = e^{iv_j t} S_n(P_j, t);$$

hence $\Sigma \hat{f}(j) e^{ijt}$ diverges on E.

Conversely, assume that E is a set of divergence for B. By remark (c.2) there exists a monotonic sequence $\omega_n \to \infty$ and a function $f \in B$ such that $S_n(f,t) > \omega_n$ infinitely often for every $t \in E$. We now pick a sequence of integers $\{\lambda_j\}$ such that

(c.6) $$\|f - \sigma_{\lambda_j}(f)\|_B < 2^{-j}$$

and then integers μ_j such that

(c.7) $$\omega_{\mu_j} > 2 \sup_t S^*(\sigma_{\lambda_j}(f), t)$$

and write $P_j = V_{\mu_{j+1}} * (f - \sigma_{\lambda_j}(f))$

where * denotes convolution and where V_μ denotes

de la Vallee Poussin's kernel $\left(2K_{2\mu+1}(t) - K_{\mu}(t)\right)$ where $k_{\mu}(t)$ is Fejer's kernel. It follows immediately from (c.6) that $\Sigma \|P_j\|_B < \infty$ If $t \varepsilon E$ and n is an integer such that $|S_n(f,t)| > \omega_n$, then for some j, $\mu_j < n \leq \mu_{j+1}$ and

$$S_n(P_j, t) = S_n(f - \sigma_{\lambda_j}(f), t) = S_n(f,t) - S_n(\sigma_{\lambda_j}(f),t).$$

Hence, by (c.7), $|S_n(P_j,t)| > \frac{1}{2}\omega_n$, and (c.5) follows.

Theorem: Assume (c.4). Let E_j be sets of divergence for B, $j = 1,2, \ldots$; then $E = \bigcup_{j=1}^{\infty} E_j$ is a set of divergence for B.

PROOF: Let $\{P_n^j\}$ be the sequence of polynomials corresponding to E_j. Omitting a finite number of terms does not change (c.5), but permits us to assume $\Sigma_{j,n}\|P_n^j\|_B < \infty$ which shows, by the lemma, that E is a set of divergence for B.

(C.4) We turn now to examine the sets of divergence for $B = C(T)$, the space of all continuous 2π-periodic functions.

Lemma: Let E be a union of a finite number of intervals on T; denote the measure of E by δ. There exists a trigonometric polynomial ϕ such that

$$S^*(\phi,t) > \frac{1}{2\pi}\log\left(\frac{1}{3\delta}\right) \text{ on } E$$

(c.8)

$$\|\phi\|_{\infty} \leq 1.$$

PROOF: It is convenient to identify T with the unit circumference $\{z; |z| = 1\}$. Let I be a (small) interval on T, $I = \{e^{it}, |t - t_o| \leq \varepsilon\}$ the function $\Psi_I = (1 + \varepsilon - ze^{-ito})^{-1}$ has a positive real part throughout the unit disc, its real part is larger than $1/3\varepsilon$ on I, and its value at the origin (z = 0) is $(1 + \varepsilon)^{-1}$. We now write $E \subset \bigcup_1^N I_j$, the I_j being small intervals of equal length 2ε such that $N\varepsilon < \delta$, and consider the function

$$\Psi(z) = \frac{1 + \varepsilon}{N} \Sigma \Psi_{I_j}(x).$$

Ψ has the following properties:

$$\begin{aligned} &\mathrm{Re}(\Psi(x)) > 0 \quad \text{for } |z| \leq 1 \\ (c.9) \qquad &\Psi(0) = 1 \\ &|\Psi(z)| \geq \mathrm{Re}(\Psi(z)) > \frac{1}{3N\varepsilon} > \frac{1}{3\delta} \quad \text{on } E. \end{aligned}$$

The function $\log \Psi$ which takes the value zero at z = 0 is holomorphic in a neighborhood of $\{z; |z| \leq 1\}$ and has the properties

$$(c.10) \qquad \begin{aligned} &|\mathrm{Im}(\log \Psi(z))| < \pi && \text{on } T \\ &|\log \Psi(z)| > \log(3\delta)^{-1} && \text{on } E. \end{aligned}$$

Since the Taylor series of $\log \Psi$ converges uniformly on T, we can take a partial sum $\Phi(z) = \Sigma_1^M a_n z^n$ of that series such that (3.10) is valid for Φ in place of $\log \Psi$. We can now put

$$\phi(t) = \frac{1}{\pi} e^{-iMt} \mathrm{Im}(\Phi(e^{it})) = \frac{1}{2\pi i} e^{-iMt} \left(\sum_1^M a_n e^{int} - \sum_1^M a_n e^{-int} \right)$$

and notice that

$$|S_M(\phi, t) = \frac{1}{2\pi}|\Phi(e^{it})|.$$

Theorem: Every set of measure zero is a set of divergence for C(T).

PROOF: If E is a set of measure zero, it can be covered by a union $\bigcup I_n$, the I_n being intervals of length $|I_n|$ such that $\Sigma |I_n| < 1$ and such that every $t \in E$ belongs to infinitely many I_n's, Grouping finite sets of intervals we can cover E infinitely often by $\bigcup E_n$ such that every E_n is a finite union of intervals and such that $|E_n| < e^{-2^n}$. Let ϕ_n be a polynomial satisfying (c.8) for $E = E_n$ and put $P_n = n^{-2}\phi_n$, We clearly have $\Sigma \|P_n\|_\infty < \infty$ and $S^*(P_n, t) > 2^{n-1}/2\pi n^2$ on E_n. Since every $t \in E$ belongs to infinitely many E_n's our theorem follows from lemma (c.3).

(C.5) Theorem: Let B be a homogeneous Banach space on T satisfying the condition (c.4). Assume $B \supset C(T)$; then either T is a set of divergence for B or the sets of divergence for B are precisely the sets of measure zero.

PROOF: By theorem (c.4) it is clear that every set of measure zero is a set of divergence for B. All that we have to show in order to complete the proof is that, if some set of positive measure is a set of divergence for B, then T is a set of divergence for B.

Assume that E is a set of divergence of positive measure. For $\alpha \in T$ denote by E_α the translate of E by α; E_α is clearly a set of divergence for B. Let $\{\alpha_n\}$ be the sequence of all rational multiples of 2π and put $\tilde{E} = \bigcup E_{\alpha_n}$. By theorem c.2 $\tilde{E}$ is a set of divergence, and we claim that $T-\tilde{E}$ is a set of measure zero. In order to prove that, we denote by χ the characteristic function of $\tilde{E}$ and notice that

$$\chi(t - \alpha_n) = \chi(t) \qquad \text{for all } t \text{ and } \alpha_n .$$

This means

$$\sum_j \hat{\chi}(j) e^{-i\alpha_n j} e^{ijt} = \sum_j \hat{\chi}(j) e^{ijt}$$

or

$$\hat{\chi}(j) e^{-i\alpha_n j} = \hat{\chi}(j) \qquad (\text{all } \alpha_n)$$

If $j \neq 0$, this implies $\hat{\chi}(j) = 0$; hence $\chi(t) =$ constant almost everywhere and, since χ is a characteristic function, this implies that the measure of $\tilde{E}$ is either zero or 2π. Since $\tilde{E} \supset E$, $\tilde{E}$ is almost all of T. Now $T - \tilde{E}$ is a set of divergence (being of measure zero) and $\tilde{E}$ is a set of divergence, hence T is a set of divergence.

(C.6) Thus, for spaces B satisfying the conditions of theorem c.5, and in particular for $B = L^p(T)$, $1 \le p < \infty$, or $B = C(T)$, either there exists a function $f \in B$ whose Fourier series diverges everywhere, or the Fourier series of every $f \in B$ converges almost everywhere.

Theorem: There exists a Fourier series diverging everywhere.

PROOF: For arbitrary $\kappa > 0$ we shall describe a positive measure μ_κ of total mass one having the property that for almost all $t \in T$.

(c.11) $$S^*(\mu_\kappa, t) = \sup_n |S_n(\mu_\kappa, t)| > \kappa .$$

Assume for the moment that such μ_κ exist; it follows from (c.11) that there exists an integer N_κ and a set E_κ of (normalized Lebesgue) measure greater than $1 - 1/\kappa$, such that for $t \in E_\kappa$

(c.12) $$\sup_{n<N\kappa} |S_n(\mu_\kappa, t)| > \kappa .$$

If we write now $\phi_\kappa = \mu_\kappa * V_{N\kappa}$ ($V_{N\kappa}$ being de la Vallee Poussin's kernel), then ϕ_κ is a trigonometric polynomial, $\|\phi_\kappa\|_{L^1} \leq 3$ and

$$S^*(\phi_\kappa, t) \geq \sup_{n<N_\kappa} |S_n(\phi_\kappa, t)| = \sup_{n<N\kappa} |S_n(\mu_\kappa, t)| > \kappa \text{ on } E_\kappa .$$

Applying the lemma c.3 with $P_j = j^{-2} \phi_{2^j}$ we obtain that $E = \bigcap_m \bigcup_{m \leq j} E_{2^j}$ is a set of divergence for $L^1(T)$. Since E is almost all T, Kolmogorov's theorem would follow from theorem c.5.

The description of the measures μ_κ is very simple; however, for the proof that (c.11) holds for almost all $t \in T$, we shall need the following very important theorem of Kronecker.

Theorem (Kronecker): Let $x_1, \ldots, x_N$ be real numbers such that $x_1, \ldots, x_N, \pi$ are linearly independent over the field of rational numbers. Let $\varepsilon > 0$ and $\alpha_1, \ldots, \alpha_N$ be real numbers, then there exists an integer n such that

$$|e^{inx_j} - e^{i\alpha_j}| < \varepsilon, \qquad j = 1, \ldots, N.$$

We construct now the measures μ_κ as follows: let N be an integer, let x_j, $j = 1, \ldots, N$ be real numbers such that $x_1, \ldots, x_N, \pi$ are linearly independent over the rationals and such that $|x_j - (2\pi j/N)| < 1/N^2$, and let μ be the measure $1/N \, \Sigma\delta_{x_j}$

For $t \in T$ we have

$$S_n(\mu,t) = \int D_n(t - x)d\mu(x) = \frac{1}{N} \sum_1^N D_n(t - x_j) =$$

$$= \frac{1}{N} \sum_1^N \frac{\sin(n + \frac{1}{2})(t-x_j)}{\sin \frac{1}{2}(t-x_j)}$$

For almost all $t \in T$, the numbers $t - x_1, \ldots, t - x_n, \pi$ are linearly independent over the rationals. By Kronecker's theorem there exist, for each such t, integers n such that

$$\left| e^{i(n+\frac{1}{2})(t-x_j)} - i \operatorname{sgn}\left(\sin \frac{t-x_j}{2}\right)\right| < \frac{1}{2}, \quad j = 1, \ldots, N;$$

hence

$$\frac{\sin(n+\frac{1}{2})(t-x_j)}{\sin \frac{1}{2}(t-x_j)} > \frac{1}{2}\left|\sin \frac{t-x_j}{2}\right|^{-1} \quad \text{for all } j.$$

It follows that

(c.13) $$S_n(\mu,t) > \frac{1}{2N} \sum_{j=1}^{N} \left| \sin \frac{t-x_j}{2} \right|^{-1}$$

and since the x_j's are so close to the roots of unity of order N, the sum in (c.13) is bounded below by

$$\frac{1}{2} \int_{1/N}^{\pi} |\sin t/2|^{-1} dt > \log N > \kappa$$, provided we take N large enough.

(C.7) In [4] Y.M. Chen modifies the classical construction of Kolmogorov's function that is given in [1] section 17 in Vol. 1 to obtain a function of the class $L(\log^+ \log^+ L)^{1-\varepsilon}$ for any $\varepsilon > 0$ whose Fourier series diverges almost everywhere.

Also, all the material in this appendix has been taken verbatim with the author's permission from [11].

BIBLIOGRAPHY

1. Bary, N.A. Treatise on Trigonometric Series, Vols. 1 and 2. Pergamon Press, Inc., New York (1964).

2. Carleson, L. On convergence and growth of partial sums of Fourier series. Acta Math. 116 (1966), 135-157.

3. Carleson, L. Sur la convergence et l'ordre de grandeur des sommes partielles des series de Fourier. (unpublished Marseille notes)

4. Chen, Y.M. An almost everywhere divergent Fourier series of the class $L(\log^+\log^+ L)^{1-\varepsilon}$. J. London Math. Soc., 44 (1969), 643-654.

5. Edwards, R.E. Fourier Series a Modern Introduction, Vols. 1 and 2. Holt, Rinehart and Winston, Inc., New York (1967).

6. Halmos, P.R. Measure Theory. Van Nostrand, New York, (1950).

7. Hewitt, E. and Stromberg, K. Real and Abstract Analysis, Springer-Verlag, Berlin (1965).

8. Hunt, R.A. On the convergence of Fourier series. (unpublished Chicago notes)

9. Hunt, R.A. On the convergence of Fourier series. Orthogonal Expansions and Their Continuous Analogues (Proc. Conf. Edwardsville, Ill. (1967)) pp. 235-255. Southern Illinois Univ. Press, Carbondale, Ill., (1968).

10. Hunt, R.A. On $L(p,q)$ spaces, Enseignment Math. 12 (1966), 249-276.

11. Katznelson, Y. An Introduction to Harmonic Analysis. John Wiley and Sons, New York (1968).

12. Mátté, A. The convergence of Fourier series of square integrable functions. Matematikai Lapok 18 (1967), 195-242. (in Hungarian)

13. Rudin, W. *Real and Complex Analysis,* McGraw-Hill New York (1966).

14. Stein, E.M. and Weiss, G. *An extension of a theorem of Marcinikiewicz and some of its applications.* J. Math. Mech. 8 (1959), 263-284.

15. Titchmarsh, E.C. *An Introduction to the Theory of Fourier Integrals.* Oxford University Press, New York (1948).

16. Zygmund, A. *Trigonometric Series,* Vols. 1 and 2. Cambridge University Press, New York (1959).

Lecture Notes in Mathematics

Vol. 85: P. Cartier et D. Foata, Problèmes combinatoires de commutation et réarrangements. IV, 88 pages. 1969. DM 8,– / $ 2.20

Vol. 86: Category Theory, Homology Theory and their Applications I. Edited by P. Hilton. VI, 216 pages. 1969. DM 16,– / $ 4.40

Vol. 87: M. Tierney, Categorical Constructions in Stable Homotopy Theory. IV, 65 pages. 1969. DM 6,– / $ 1.70

Vol. 88: Séminaire de Probabilités III. IV, 229 pages. 1969. DM 18,– / $ 5.00

Vol. 89: Probability and Information Theory. Edited by M. Behara, K. Krickeberg and J. Wolfowitz. IV, 256 pages. 1969. DM 18,– / $ 5.00

Vol. 90: N. P. Bhatia and O. Hajek, Local Semi-Dynamical Systems. II, 157 pages. 1969. DM 14,– / $ 3.90

Vol. 91: N. N. Janenko, Die Zwischenschrittmethode zur Lösung mehrdimensionaler Probleme der mathematischen Physik. VIII, 194 Seiten. 1969. DM 16,80 / $ 4.70

Vol. 92: Category Theory, Homology Theory and their Applications II. Edited by P. Hilton. V, 308 pages. 1969. DM 20,– / $ 5.50

Vol. 93: K. R. Parthasarathy, Multipliers on Locally Compact Groups. III, 54 pages. 1969. DM 5,60 / $ 1.60

Vol. 94: M. Machover and J. Hirschfeld, Lectures on Non-Standard Analysis. VI, 79 pages. 1969. DM 6,– / $ 1.70

Vol. 95: A. S. Troelstra, Principles of Intuitionism. II, 111 pages. 1969. DM 10,– / $ 2.80

Vol. 96: H.-B. Brinkmann und D. Puppe, Abelsche und exakte Kategorien, Korrespondenzen. V, 141 Seiten. 1969. DM 10,– / $ 2.80

Vol. 97: S. O. Chase and M. E. Sweedler, Hopf Algebras and Galois theory. II, 133 pages. 1969. DM 10,– / $ 2.80

Vol. 98: M. Heins, Hardy Classes on Riemann Surfaces. III, 106 pages. 1969. DM 10,– / $ 2.80

Vol. 99: Category Theory, Homology Theory and their Applications III. Edited by P. Hilton. IV, 489 pages. 1969. DM 24,– / $ 6.60

Vol. 100: M. Artin and B. Mazur, Etale Homotopy. II, 196 Seiten. 1969. DM 12,– / $ 3.30

Vol. 101: G. P. Szegö et G. Treccani, Semigruppi di Trasformazioni Multivoche. VI, 177 pages. 1969. DM 14,– / $ 3.90

Vol. 102: F. Stummel, Rand- und Eigenwertaufgaben in Sobolewschen Räumen. VIII, 386 Seiten. 1969. DM 20,– / $ 5.50

Vol. 103: Lectures in Modern Analysis and Applications I. Edited by C. T. Taam. VII, 162 pages. 1969. DM 12,– / $ 3.30

Vol. 104: G. H. Pimbley, Jr., Eigenfunction Branches of Nonlinear Operators and their Bifurcations. II, 128 pages. 1969. DM 10,– / $ 2.80

Vol. 105: R. Larsen, The Multiplier Problem. VII, 284 pages. 1969. DM 18,– / $ 5.00

Vol. 106: Reports of the Midwest Category Seminar III. Edited by S. Mac Lane. III, 247 pages. 1969. DM 16,– / $ 4.40

Vol. 107: A. Peyerimhoff, Lectures on Summability. III, 111 pages. 1969. DM 8,– / $ 2.20

Vol. 108: Algebraic K-Theory and its Geometric Applications. Edited by R. M.F. Moss and C. B. Thomas. IV, 86 pages. 1969. DM 6,– / $ 1.70

Vol. 109: Conference on the Numerical Solution of Differential Equations. Edited by J. Ll. Morris. VI, 275 pages. 1969. DM 18,– / $ 5.00

Vol. 110: The Many Facets of Graph Theory. Edited by G. Chartrand and S. F. Kapoor. VIII, 290 pages. 1969. DM 18,– / $ 5.00

Vol. 111: K. H. Mayer, Relationen zwischen charakteristischen Zahlen. III, 99 Seiten. 1969. DM 8,– / $ 2.20

Vol. 112: Colloquium on Methods of Optimization. Edited by N. N. Moiseev. IV, 293 pages. 1970. DM 18,– / $ 5.00

Vol. 113: R. Wille, Kongruenzklassengeometrien. III, 99 Seiten. 1970. DM 8,– / $ 2.20

Vol. 114: H. Jacquet and R. P. Langlands, Automorphic Forms on GL (2). VII, 548 pages. 1970. DM 24,– / $ 6.60

Vol. 115: K. H. Roggenkamp and V. Huber-Dyson, Lattices over Orders I. XIX, 290 pages. 1970. DM 18,– / $ 5.00

Vol. 116: Séminaire Pierre Lelong (Analyse) Année 1969. IV, 195 pages. 1970. DM 14,– / $ 3.90

Vol. 117: Y. Meyer, Nombres de Pisot, Nombres de Salem et Analyse Harmonique. 63 pages. 1970. DM 6.– / $ 1.70

Vol. 118: Proceedings of the 15th Scandinavian Congress, Oslo 1968. Edited by K. E. Aubert and W. Ljunggren. IV, 162 pages. 1970. DM 12,– / $ 3.30

Vol. 119: M. Raynaud, Faisceaux amples sur les schémas en groupes et les espaces homogènes. III, 219 pages. 1970. DM 14,– / $ 3.90

Vol. 120: D. Siefkes, Büchi's Monadic Second Order Successor Arithmetic. XII, 130 Seiten. 1970. DM 12,– / $ 3.30

Vol. 121: H. S. Bear, Lectures on Gleason Parts. III, 47 pages. 1970. DM 6,– / $ 1.70

Vol. 122: H. Zieschang, E. Vogt und H.-D. Coldewey, Flächen und ebene diskontinuierliche Gruppen. VIII, 203 Seiten. 1970. DM 16,– / $ 4.40

Vol. 123: A. V. Jategaonkar, Left Principal Ideal Rings. VI, 145 pages. 1970. DM 12,– / $ 3.30

Vol. 124: Séminare de Probabilités IV. Edited by P. A. Meyer. IV, 282 pages. 1970. DM 20,– / $ 5.50

Vol. 125: Symposium on Automatic Demonstration. V, 310 pages. 1970. DM 20,– / $ 5.50

Vol. 126: P. Schapira, Théorie des Hyperfonctions. XI, 157 pages. 1970. DM 14,– / $ 3.90

Vol. 127: I. Stewart, Lie Algebras. IV, 97 pages. 1970. DM 10,– / $ 2.80

Vol. 128: M. Takesaki, Tomita's Theory of Modular Hilbert Algebras and its Applications. II, 123 pages. 1970. DM 10,– / $ 2.80

Vol. 129: K. H. Hofmann, The Duality of Compact Semigroups and C*- Bigebras. XII, 142 pages. 1970. DM 14,– / $ 3.90

Vol. 130: F. Lorenz, Quadratische Formen über Körpern. II, 77 Seiten. 1970. DM 8,– / $ 2.20

Vol. 131: A Borel et al., Seminar on Algebraic Groups and Related Finite Groups. VII, 321 pages. 1970. DM 22,– / $ 6.10

Vol. 132: Symposium on Optimization. III, 348 pages. 1970. DM 22,– / $ 6.10

Vol. 133: F. Topsøe, Topology and Measure. XIV, 79 pages. 1970. DM 8,– / $ 2.20

Vol. 134: L. Smith, Lectures on the Eilenberg-Moore Spectral Sequence. VII, 142 pages. 1970. DM 14,– / $ 3.90

Vol. 135: W. Stoll, Value Distribution of Holomorphic Maps into Compact Complex Manifolds. II, 267 pages. 1970. DM 18,– / $ 5.00

Vol. 136: M. Karoubi et al., Séminaire Heidelberg-Saarbrücken-Strasbuorg sur la K-Théorie. IV, 264 pages. 1970. DM 18,– / $ 5.00

Vol. 137: Reports of the Midwest Category Seminar IV. Edited by S. MacLane. III, 139 pages. 1970. DM 12,– / $ 3.30

Vol. 138: D. Foata et M. Schützenberger, Théorie Géométrique des Polynômes Eulériens. V, 94 pages. 1970. DM 10,– / $ 2.80

Vol. 139: A. Badrikian, Séminaire sur les Fonctions Aléatoires Linéaires et les Mesures Cylindriques. VII, 221 pages. 1970. DM 18,– / $ 5.00

Vol. 140: Lectures in Modern Analysis and Applications II. Edited by C. T. Taam. VI, 119 pages. 1970. DM 10,– / $ 2.80

Vol. 141: G. Jameson, Ordered Linear Spaces. XV, 194 pages. 1970. DM 16,– / $ 4.40

Vol. 142: K. W. Roggenkamp, Lattices over Orders II. V, 388 pages. 1970. DM 22,– / $ 6.10

Vol. 143: K. W. Gruenberg, Cohomological Topics in Group Theory. XIV, 275 pages. 1970. DM 20,– / $ 5.50

Vol. 144: Seminar on Differential Equations and Dynamical Systems, II. Edited by J. A. Yorke. VIII, 268 pages. 1970. DM 20,– / $ 5.50

Vol. 145: E. J. Dubuc, Kan Extensions in Enriched Category Theory. XVI, 173 pages. 1970. DM 16,– / $ 4.40

Vol. 146: A. B. Altman and S. Kleiman, Introduction to Grothendieck Duality Theory. II, 192 pages. 1970. DM 18,– / $ 5.00

Vol. 147: D. E. Dobbs, Cech Cohomological Dimensions for Commutative Rings. VI, 176 pages. 1970. DM 16,– / $ 4.40

Vol. 148: R. Azencott, Espaces de Poisson des Groupes Localement Compacts. IX, 141 pages. 1970. DM 14,– / $ 3.90

Vol. 149: R. G. Swan and E. G. Evans, K-Theory of Finite Groups and Orders. IV, 237 pages. 1970. DM 20,– / $ 5.50

Vol. 150: Heyer, Dualität lokalkompakter Gruppen. XIII, 372 Seiten. 1970. DM 20,– / $ 5.50

Vol. 151: M. Demazure et A. Grothendieck, Schémas en Groupes I. (SGA 3). XV, 562 pages. 1970. DM 24,– / $ 6.60

Vol. 152: M. Demazure et A. Grothendieck, Schémas en Groupes II. (SGA 3). IX, 654 pages. 1970. DM 24,– / $ 6.60

Vol. 153: M. Demazure et A. Grothendieck, Schémas en Groupes III. (SGA 3). VIII, 529 pages. 1970. DM 24,– / $ 6.60

Vol. 154: A. Lascoux et M. Berger, Variétés Kähleriennes Compactes. VII, 83 pages. 1970. DM 8,– / $ 2.20

Lecture Notes in Mathematics – Lecture Notes in Physics

Vol. 155: Several Complex Variables I, Maryland 1970. Edited by J. Horváth. IV, 214 pages. 1970. DM 18,– / $ 5.00

Vol. 156: R. Hartshorne, Ample Subvarieties of Algebraic Varieties. XIV, 256 pages. 1970. DM 20,– / $ 5.50

Vol. 157: T. tom Dieck, K. H. Kamps und D. Puppe, Homotopietheorie. VI, 265 Seiten. 1970. DM 20,– / $ 5.50

Vol. 158: T. G. Ostrom, Finite Translation Planes. IV. 112 pages. 1970. DM 10,– / $ 2.80

Vol. 159: R. Ansorge und R. Hass. Konvergenz von Differenzenverfahren für lineare und nichtlineare Anfangswertaufgaben. VIII, 145 Seiten. 1970. DM 14,– / $ 3.90

Vol. 160: L. Sucheston, Constributions to Ergodic Theory and Probability. VII, 277 pages. 1970. DM 20,– / $ 5.50

Vol. 161: J. Stasheff, H-Spaces from a Homotopy Point of View. VI, 95 pages. 1970. DM 10,– / $ 2.80

Vol. 162: Harish-Chandra and van Dijk, Harmonic Analysis on Reductive p-adic Groups. IV, 125 pages. 1970. DM 12,– / $ 3.30

Vol. 163: P. Deligne, Equations Différentielles à Points Singuliers Reguliers. III, 133 pages. 1970. DM 12,– / $ 3.30

Vol. 164: J. P. Ferrier, Seminaire sur les Algebres Complètes. II, 69 pages. 1970. DM 8,– / $ 2.20

Vol. 165: J. M. Cohen, Stable Homotopy. V, 194 pages. 1970. DM 16,– / $ 4.40

Vol. 166: A. J. Silberger, PGL_2 over the p-adics: its Representations, Spherical Functions, and Fourier Analysis. VII, 202 pages. 1970. DM 18,– / $ 5.00

Vol. 167: Lavrentiev, Romanov and Vasiliev, Multidimensional Inverse Problems for Differential Equations. V, 59 pages. 1970. DM 10,– / $ 2.80

Vol. 168: F. P. Peterson, The Steenrod Algebra and its Applications: A conference to Celebrate N. E. Steenrod's Sixtieth Birthday. VII, 317 pages. 1970. DM 22,– / $ 6.10

Vol. 169: M. Raynaud, Anneaux Locaux Henséliens. V, 129 pages. 1970. DM 12,– / $ 3.30

Vol. 170: Lectures in Modern Analysis and Applications III. Edited by C. T. Taam. VI, 213 pages. 1970. DM 18,– / $ 5.00.

Vol. 171: Set-Valued Mappings, Selections and Topological Properties of 2^X. Edited by W. M. Fleischman. X, 110 pages. 1970. DM 12,– / $ 3.30

Vol. 172: Y.-T. Siu and G. Trautmann, Gap-Sheaves and Extension of Coherent Analytic Subsheaves. V, 172 pages. 1971. DM 16,– / $ 4.40

Vol. 173: J. N. Mordeson and B. Vinograde, Structure of Arbitrary Purely Inseparable Extension Fields. IV, 138 pages. 1970. DM 14,– / $ 3.90.

Vol. 174: B. Iversen, Linear Determinants with Applications to the Picard Scheme of a Family of Algebraic Curves. VI, 69 pages. 1970. DM 8,– / $ 2.20.

Vol. 175: M. Brelot, On Topologies and Boundaries in Potential Theory. VI, 176 pages. 1971. DM 18,– / $ 5.00

Vol. 176: H. Popp, Fundamentalgruppen algebraischer Mannigfaltigkeiten. IV, 154 Seiten. 1970. DM 16,– / $ 4.40

Vol. 177: J. Lambek, Torsion Theories, Additive Semantics and Rings of Quotients. VI, 94 pages. 1971. DM 12,– / $ 3.30

Vol. 178: Th. Bröcker und T. tom Dieck, Kobordismentheorie. XVI, 191 Seiten. 1970. DM 18,– / $ 5.00

Vol. 179: Seminaire Bourbaki – vol. 1968/69. Exposés 347-363. IV. 295 pages. 1971. DM 22,– / $ 6.10

Vol. 180: Séminaire Bourbaki – vol. 1969/70. Exposés 364-381. IV, 310 pages. 1971. DM 22,– / $ 6.10

Vol. 181: F. DeMeyer and E. Ingraham, Separable Algebras over Commutative Rings. V, 157 pages. 1971. DM 16.– / $ 4.40

Vol. 182: L. D. Baumert. Cyclic Difference Sets. VI, 166 pages. 1971. DM 16,– / $ 4.40

Vol. 183: Analytic Theory of Differential Equations. Edited by P. F. Hsieh and A. W. J. Stoddart. VI, 225 pages. 1971. DM 20,– / $ 5.50

Vol. 184: Symposium on Several Complex Variables, Park City, Utah, 1970. Edited by R. M. Brooks. V, 234 pages. 1971. DM 20,– / $ 5.50

Vol. 185: Several Complex Variables II, Maryland 1970. Edited by J. Horváth. III, 287 pages. 1971. DM 24,– / $ 6.60

Vol. 186: Recent Trends in Graph Theory. Edited by M. Capobianco/ J. B. Frechen/M. Krolik. VI, 219 pages. 1971. DM 18.– / $ 5.00

Vol. 187: H. S. Shapiro, Topics in Approximation Theory. VIII, 275 pages. 1971. DM 22,– / $ 6.10

Vol. 188: Symposium on Semantics of Algorithmic Languages. Edited by E. Engeler. VI, 372 pages. 1971. DM 26,– / $ 7.20

Vol. 189: A. Weil, Dirichlet Series and Automorphic Forms. V, 164 pages. 1971. DM 16,– / $ 4.40

Vol. 190: Martingales. A Report on a Meeting at Oberwolfach, May 17-23, 1970. Edited by H. Dinges. V, 75 pages. 1971. DM 12,– / $ 3.30

Vol. 191: Séminaire de Probabilités V. Edited by P. A. Meyer. IV, 372 pages. 1971. DM 26,– / $ 7.20

Vol. 192: Proceedings of Liverpool Singularities – Symposium I. Edited by C. T. C. Wall. V, 319 pages. 1971. DM 24,– / $ 6.60

Vol. 193: Symposium on the Theory of Numerical Analysis. Edited by J. Ll. Morris. VI, 152 pages. 1971. DM 16,– / $ 4.40

Vol. 194: M. Berger, P. Gauduchon et E. Mazet. Le Spectre d'une Variété Riemannienne. VII, 251 pages. 1971. DM 22,– / $ 6.10

Vol. 195: Reports of the Midwest Category Seminar V. Edited by J.W. Gray and S. Mac Lane. III, 255 pages. 1971. DM 22,– / $ 6.10

Vol. 196: H-spaces – Neuchâtel (Suisse)- Août 1970. Edited by F. Sigrist. V, 156 pages. 1971. DM 16,– / $ 4.40

Vol. 197: Manifolds – Amsterdam 1970. Edited by N. H. Kuiper. V, 231 pages. 1971. DM 20,– / $ 5.50

Vol. 198: M. Hervé, Analytic and Plurisubharmonic Functions in Finite and Infinite Dimensional Spaces. VI, 90 pages. 1971. DM 16.– / $ 4.40

Vol. 199: Ch. J. Mozzochi, On the Pointwise Convergence of Fourier Series. VII, 87 pages. 1971. DM 16,– / $ 4.40

Lecture Notes in Physics

Vol. 1: J. C. Erdmann, Wärmeleitung in Kristallen, theoretische Grundlagen und fortgeschrittene experimentelle Methoden. II, 283 Seiten. 1969. DM 20,– / $ 5.50

Vol. 2: K. Hepp, Théorie de la renormalisation. III, 215 pages. 1969. DM 18,– / $ 5.00

Vol. 3: A. Martin, Scattering Theory: Unitarity, Analytic and Crossing. IV, 125 pages. 1969. DM 14,– / $ 3.90

Vol. 4: G. Ludwig, Deutung des Begriffs physikalische Theorie und axiomatische Grundlegung der Hilbertraumstruktur der Quantenmechanik durch Hauptsätze des Messens. XI, 469 Seiten. 1970. DM 28,– / $ 7.70

Vol. 5: M. Schaaf, The Reduction of the Product of Two Irreducible Unitary Representations of the Proper Orthochronous Quantummechanical Poincaré Group. IV, 120 pages. 1970. DM 14,– / $ 3.90

Vol. 6: Group Representations in Mathematics and Physics. Edited by V. Bargmann. V, 340 pages. 1970. DM 24,– / $ 6.60

Vol. 7: R. Balescu, J. L. Lebowitz, I. Prigogine, P. Résibois, Z. W. Salsburg, Lectures in Statistical Physics. V, 181 pages. 1971. DM 18,– / $ 5.00

Vol. 8: Proceedings of the Second International Conference on Numerical Methods in Fluid Dynamics. Edited by M. Holt. IX, 462 pages. 1971. DM 28,– / $ 7.70

Beschaffenheit der Manuskripte

Die Manuskripte werden photomechanisch vervielfältigt; sie müssen daher in sauberer Schreibmaschinenschrift geschrieben sein. Handschriftliche Formeln bitte nur mit schwarzer Tusche eintragen. Notwendige Korrekturen sind bei dem bereits geschriebenen Text entweder durch Überkleben des alten Textes vorzunehmen oder aber müssen die zu korrigierenden Stellen mit weißem Korrekturlack abgedeckt werden. Falls das Manuskript oder Teile desselben neu geschrieben werden müssen, ist der Verlag bereit, dem Autor bei Erscheinen seines Bandes einen angemessenen Betrag zu zahlen. Die Autoren erhalten 75 Freiexemplare.

Zur Erreichung eines möglichst optimalen Reproduktionsergebnisses ist es erwünscht, daß bei der vorgesehenen Verkleinerung der Manuskripte der Text auf einer Seite in der Breite möglichst 18 cm und in der Höhe 26,5 cm nicht überschreitet. Entsprechende Satzspiegelvordrucke werden vom Verlag gern auf Anforderung zur Verfügung gestellt.

Manuskripte, in englischer, deutscher oder französischer Sprache abgefaßt, nimmt Prof. Dr. A. Dold, Mathematisches Institut der Universität Heidelberg, Tiergartenstraße oder Prof. Dr. B. Eckmann, Eidgenössische Technische Hochschule, Zürich, entgegen.

Cette série a pour but de donner des informations rapides, de niveau élevé, sur des développements récents en mathématiques, aussi bien dans la recherche que dans l'enseignement supérieur. On prévoit de publier

1. des versions préliminaires de travaux originaux et de monographies
2. des cours spéciaux portant sur un domaine nouveau ou sur des aspects nouveaux de domaines classiques
3. des rapports de séminaires
4. des conférences faites à des congrès ou à des colloquiums

En outre il est prévu de publier dans cette série, si la demande le justifie, des rapports de séminaires et des cours multicopiés ailleurs mais déjà épuisés.

Dans l'intérêt d'une diffusion rapide, les contributions auront souvent un caractère provisoire; le cas échéant, les démonstrations ne seront données que dans les grandes lignes. Les travaux présentés pourront également paraître ailleurs. Une réserve suffisante d'exemplaires sera toujours disponible. En permettant aux personnes intéressées d'être informées plus rapidement, les éditeurs Springer espèrent, par cette série de »prépublications«, rendre d'appréciables services aux instituts de mathématiques. Les annonces dans les revues spécialisées, les inscriptions aux catalogues et les copyrights rendront plus facile aux bibliothèques la tâche de réunir une documentation complète.

Présentation des manuscrits

Les manuscrits, étant reproduits par procédé photomécanique, doivent être soigneusement dactylographiés. Il est recommandé d'écrire à l'encre de Chine noire les formules non dactylographiées. Les corrections nécessaires doivent être effectuées soit par collage du nouveau texte sur l'ancien soit en recouvrant les endroits à corriger par du verni correcteur blanc.

S'il s'avère nécessaire d'écrire de nouveau le manuscrit, soit complètement, soit en partie, la maison d'édition se déclare prête à verser à l'auteur, lors de la parution du volume, le montant des frais correspondants. Les auteurs recoivent 75 exemplaires gratuits.

Pour obtenir une reproduction optimale il est désirable que le texte dactylographié sur une page ne dépasse pas 26,5 cm en hauteur et 18 cm en largeur. Sur demande la maison d'édition met à la disposition des auteurs du papier spécialement préparé.

Les manuscrits en anglais, allemand ou français peuvent être adressés au Prof. Dr. A. Dold, Mathematisches Institut der Universität Heidelberg, Tiergartenstraße ou au Prof. Dr. B. Eckmann, Eidgenössische Technische Hochschule, Zürich.

ISBN 3-540-05475-8
ISBN 0-387-05475-8